Manufacturing Success
How to Manage Your Competitive Edge

Roland Toone

Prentice Hall

New York   London   Toronto   Sydney   Tokyo   Singapore

First published 1994 by
Prentice Hall International (UK) Limited
Campus 400, Maylands Avenue
Hemel Hempstead
Hertfordshire, HP2 7EZ
A division of
Simon & Schuster International Group

© Prentice Hall International (UK) Ltd, 1994

All rights reserved. No part of this publication may be reproduced, stored in a retrieval system, or transmitted, in any form, or by any means, electronic, mechanical, photocopying, recording or otherwise, without prior permission, in writing, from the publisher.
For permission within the United States of America
contact Prentice Hall Inc., Englewood Cliffs, NJ 07632

Typeset in 10/12pt Times by
Mathematical Composition Setters Ltd, Salisbury, Wiltshire

Printed and bound in Great Britain by
T. J. Press, Padstow, Cornwall

Library of Congress Cataloging-in-Publication Data

Toone, R. P. (Roland P.), 1925–
   Manufacturing success : how to manage your competitive edge / Roland Toone.
     p.   cm. – (The Manufacturing practitioner series)
   Includes bibliographical references and index.
   ISBN 0-13-102773-5
   1. Great Britain–Manufactures–Management. 2. Great Britain–Manufactures–Management–Case studies. I. Title. II. Series
HD9731.5.T66 1994
670'.68–dc20   94-11047
                                                                                 CIP

British Library Cataloguing in Publication Data

A catalogue record for this book is available from the British Library

ISBN 0–13–102773–5 (pbk)

---

1 2 3 4 5   97 96 95 94

Manufacturing Success

The Manufacturing Practitioner Series

Stop Wasting Time: Computer-Aided Planning and Control:
 Norman Sanders
Just-in-Time Manufacturing in Perspective: Alan Harrison
Managing for Total Quality: From Deming to Taguchi and SPC:
 N. Logothetis

Contents

Preface ix

1 Britain's Position in World Manufacturing 1
1.1 The environment in which manufacturing companies compete 1
1.2 Global markets, global competition and global resourcing 9
1.3 Conclusion 12

2 The Manufacturing Environment 13
2.1 The management of successful manufacturing companies 13
2.2 Company experiences with the introduction of new tools and techniques 18
2.3 Changes in the environment within a manufacturing company 23
2.4 How a manufacturing company competes 37
2.5 Conclusion 48

3 Strategy, Integration and Focus 50
3.1 The importance of strategy, integration and focus 50
3.2 The strategy of an integrated manufacturing company 58
3.3 Developing strategy in an integrated manufacturing company 59
3.4 How to achieve focus 70
3.5 Integrating the manufacturing company 73
3.6 Conclusions 92

4 Quality — 94

4.1	Introduction	94
4.2	The costs of quality	95
4.3	Quality standards	98
4.4	The customer	100
4.5	Quality planning	104
4.6	Competitive benchmarking	106
4.7	The product and the process	113
4.8	Quality circles	117
4.9	People empowerment	119
4.10	The control of supplier quality	120
4.11	Continuous improvement	122
4.12	Conclusions	122

5 The Management of Human Resources — 125

5.1	Introduction	125
5.2	The management of the human resource during the twentieth century	127
5.3	Unions, workers and management in competitive manufacturing	132
5.4	Developing the human resource	142
5.5	Conclusion	149

6 Finance and Control — 150

6.1	The provision of finance	150
6.2	Control of costs in the business	156
6.3	Description of the techniques	159
6.4	The value chain	164
6.5	Non-financial quantitative and qualitative measures	165
6.6	Decision making	165
6.7	Conclusions	166

CASE STUDIES

Case Study 1 DSF Refractories Ltd — 171

CS1.1	Introduction	171
CS1.2	Process flow	171
CS1.3	DSF customer base	172
CS1.4	DSF business plan	173
CS1.5	Manufacturing strategy	174
CS1.6	Implementing these parts of the manufacturing strategy	177
CS1.7	Computerised planning and control	178

CS1.8	Re-organisation of production into three autonomous units	181
CS1.9	Reduction of press set-up times	182

Case Study 2 Fisher Controls Ltd 184

CS2.1	Product	184
CS2.2	History	184
CS2.3	Manufacture	184
CS2.4	Strategy	186
CS2.5	Process development	191
CS2.6	Improvements effected by the strategies	194

Case Study 3 Hepworth Building Products 195

CS3.1	The justification for manufacturing and selling concrete products	195
CS3.2	Hepworth entry strategy	196
CS3.3	Manufacturing	198
CS3.4	CIM strategy	201
CS3.5	Hardware constraints	201
CS3.6	Software constraints	201
CS3.7	Description of the CIM system	204

Case Study 4 Kodak Limited 211

CS4.1	Introduction	211
CS4.2	The manufacturing process	211
CS4.3	Bill of materials (BOM)	212
CS4.4	MRPII and change	212
CS4.5	The plan of implementation	213
CS4.6	Resource commitment	213
CS4.7	Roll film pilot	215
CS4.8	Preparation for the 135 cutover; January–April 1990	216
CS4.9	Cutover and instability: June–August 1990	216
CS4.10	135 cutover recovery phase: September–December 1990	216
CS4.11	Control and class B: January–June 1991	217
CS4.12	Class A completion: July–October 1991	217
CS4.13	MRPII benefits	217

Appendix 1	Class A MRPII user	218
Appendix 2	Time fences in MRPII	220
CS4A2.1	Introduction	220
CS4A2.2	Forecast accuracy	221
CS4A2.3	Supply leadtimes	222
CS4A2.4	Setting safety and strategic inventory	222

CS4A2.5	Managing demand over CLT	223
CS4A2.6	Cumulative leadtime, manufacturing leadtime and time zones	223
CS4A2.7	Alignment	225
CS4A2.8	Conclusion	226

Case Study 5 Rolls-Royce PLC 228

CS5.1	Manufacturing strategy in the 1980s	228
CS5.2	The advanced integrated manufacturing system (AIMS)	231
CS5.3	The development of simultaneous engineering (SE) at Rolls-Royce	234
CS5.4	Conclusion	242

Case Study 6 Stanley Tools 244

CS6.1	Introduction	244
CS6.2	Product development – the Magnum screwdriver	244
CS6.3	Manufacturing planning and control systems	247
CS6.4	Process development	249
CS6.5	Organisation in manufacturing	250

Case Study 7 The Tempered Spring Company Limited 254

CS7.1	Background	254
CS7.2	Product – engine valve springs	254
CS7.3	The total market and TSCo's share	256
CS7.4	Manufacturing	257
CS7.5	Commissioning and operating the auto-line	260
CS7.6	Success of the divisional strategy	262

List of abbreviations 264

Bibliography 266

Index 271

Preface

The object of this book is to help managers in manufacturing companies achieve success in an increasingly competitive environment. It also gives teachers and students of management a framework for analysing a successful approach to managing manufacturing.

The tools and techniques which are being offered to manufacturing managers as the way to improve company competitivity have been developed and used predominantly in the consumer durables sector. This sector manufactures automotive products, hi-fi, television, electronic products, etc.

The book examines the applicability of such tools and techniques across a much wider range of manufacturing industry. The organisation of the book is based on a matrix of case studies and chapters shown in the tabulation below.

Chapters Case Studies	The manufacturing environment	Strategy, integration and focus	Quality	Management of human resources	Finance and control
DSF Refractories					
Fisher Controls					
Hepworth					
Kodak					
Rolls-Royce					
Stanley Tools					
Tempered Spring					

Preface

Tools and techniques need to be implemented against a company-wide understanding of goals and objectives. Successful use of the tools and techniques depends on the existence of a supporting management philosophy across the company.

The case studies are presented in full and can be used as management development tools in-company and for teaching on academic courses. Extracts from the case studies are used to support the discussion in the text and to interpret choices made by companies in the face of opportunities and difficulties encountered.

The case study extracts are supplemented by illustrations drawn from published texts and company examples.

The reader is encouraged to make judgements about the use of tools and techniques:

- Why does the company require such a technique? The case study extracts show that some companies have decided that the cost of putting in a technique outweighs the benefits.
- What is the company goal in adopting a technique? Is the adoption of the technique the best way of achieving that goal? For example, adopting a system of manufacturing planning and control will only succeed if the control problems are identified, understood and simplified before choosing and installing the technique.

A theme throughout the book is the author's philosophy that strategy, integration and focus are the keys to successful management of a manufacturing company. Strategy, integration and focus enable a manufacturing company to identify the markets it can penetrate and the products required to gain competitive advantage in those markets. Products can gain competitive advantage through:

- productivity/cost;
- quality;
- delivery speed and reliability;
- flexibility.

Such products must match the needs of the customer to the needs of the company for profit and growth.

Finally the book attempts to present a coherent account of the way in which the management of manufacturing should advance. Managers are not being presented with a recipe for success but an understanding of the ingredients for success from which they can produce their own recipe. There is a history in British manufacturing companies of the enthusiastic adoption of techniques which, subsequently, have been discarded when the expected results have failed to materialise.

The case studies are an integral and invaluable part of the book. I would like to thank the companies for permission to publish the studies and the staff of those companies for their time and help in developing the studies.

Preface

- Case Study 1 – DSF Refractories Ltd.
- Case Study 2 – Fisher Controls Ltd.
- Case Study 3 – Hepworth Building Products.
- Case Study 4 – Kodak Limited.
- Case Study 5 – Rolls-Royce PLC.
- Case Study 6 – Stanley Tools.
- Case Study 7 – The Tempered Spring Company Limited.

Roland Toone

1 Britain's Position in World Manufacturing

■ 1.1 The environment in which manufacturing companies compete

1.1.1 Economic factors

Between 1970 and 1990 the gross domestic product (GDP) of the G7 countries (the world's seven most industrialised nations) increased between 60 and 146 per cent as shown in Figure 1.1. The lowest rate of growth was in the UK. Figure 1.2a shows manufacturing as a percentage of GDP for the G7 countries in 1970 and 1990. Figure 1.2b shows the decrease in manufacturing, as a percentage of the GDP, over the period.

While manufacturing has decreased in importance for all G7 countries, the decline was greatest in the UK. Here the fall was 31 per cent compared with a fall of between 13 and 22 per cent in the rest of the G7 countries.

The decline in the importance of manufacturing in the G7 economies has been caused by the growth of manufacturing in developing countries and particularly in newly industrialised nations (NICs). Figure 1.3a shows the percentage growth in GDP and Figure 1.3b shows the percentage increase in manufacturing for four NICs.

All these countries have a rapid growth in GDP and in manufacturing. For three of the countries manufacturing is increasing as a proportion of GDP. Evidently NICs are currently manufacturing a range of products in competition with existing industrialised countries. As a result global competition has increased and market shares for the G7 countries have declined.

In the G7 countries increase in service industries has compensated for the reduction in the importance of manufacturing. In the case of the UK the contribution made by North Sea oil has been a further significant factor in the growth of GDP. From Figure 1.4 it can be seen that although manufacturing in the G7 countries has become a less important part of their economies there has been real growth in manufacturing during the period.

Britain's Position in World Manufacturing

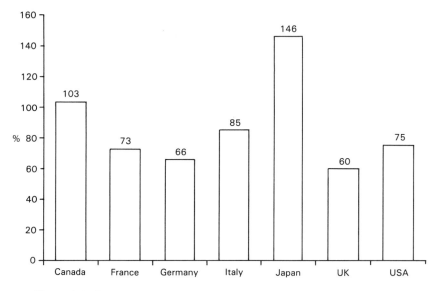

Figure 1.1 Percentage growth in gross domestic product for the G7 countries, 1970–1990 (*Source*: adapted from data published by the United Nations)

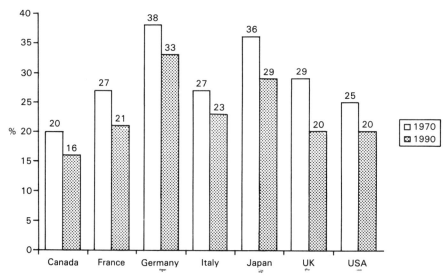

Figure 1.2a Manufacturing as a percentage of GDP for the G7 countries, 1970–1990 (*Source*: adapted from data published by the United Nations)

The competitive environment

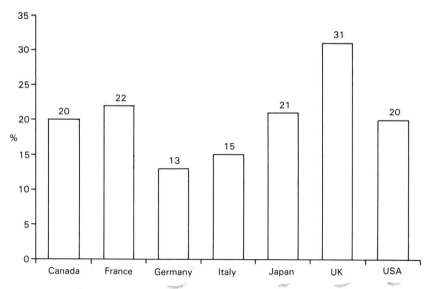

Figure 1.2b Percentage decrease of manufacturing in the GDP in the G7 countries, 1970–1990 (*Source*: adapted from data published by the United Nations)

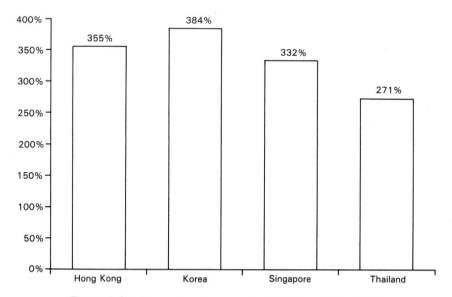

Figure 1.3a Percentage increase in GDP for selected NICs, 1970–1989 (*Source*: adapted from data published by the United Nations)

Britain's Position in World Manufacturing

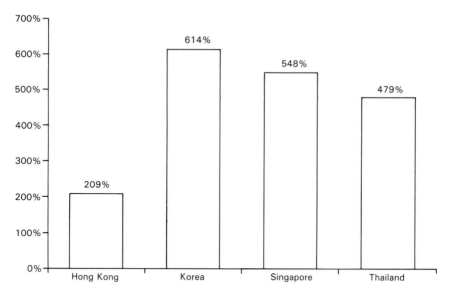

Figure 1.3b Percentage increase in manufacturing for selected NICs, 1970–1989 (*Source*: adapted from data published by the United Nations)

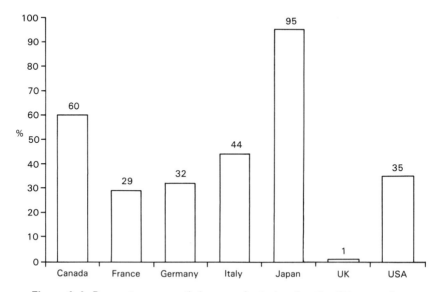

Figure 1.4 Percentage growth in manufacturing for the G7 countries, 1970–1990 (*Source*: adapted from data published by the United Nations)

The competitive environment

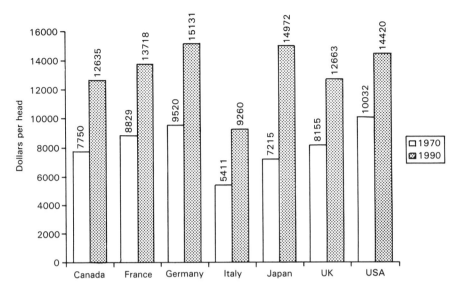

Figure 1.5a GDP per head in the G7 countries in 1980 US dollars, 1970–1990 (*Source*: adapted from data published by the United Nations)

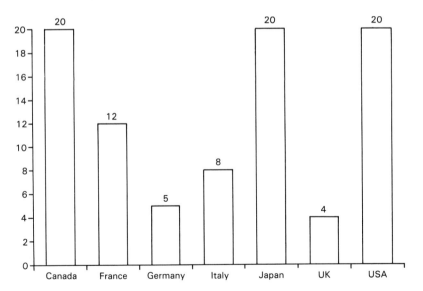

Figure 1.5b Percentage increase in GDP per head in the G7 countries, 1970–1990 (*Source*: adapted from data published by the United Nations)

Nevertheless, from the viewpoint of the UK the figures cause great concern. The growth in GDP has been the lowest in the G7 countries and less than half that of the leader, Japan. Manufacturing has grown by 1 per cent over the two decades compared with between 29 and 95 per cent in the other G7 countries. The significance of this change has been highlighted by Eltis and Fraser (1992) who comment that between 1962 and 1990 Japan's share of world exports of manufactures increased from 7 per cent to 16 per cent while for Britain the share declined from 15 per cent to 9 per cent. This change in the ability of the UK to export manufactured goods is adversely affecting the balance of payments and service industries are not able to compensate.

Completing this review of the economic position facing the UK, Figure 1.5a shows the growth in GDP per head of population in real values. Figure 1.5b shows the percentage increase in GDP per head for the G7 countries. In 1970 the UK ranked fourth in the G7 countries and had a GDP per head of $8,130, which was close to the average of $8,155. Twenty years later the UK had dropped to fifth place, and its GDP per head of $12,633 was 5 per cent lower than the average. The growth in GDP per head, Figure 1.5b, was the lowest of the G7 countries during this twenty-year period. If the trend of low growth in GDP per head continues, then relative standards of living will decline in comparison with other G7 countries. North Sea oil is a finite reserve and, when depleted, an acceleration of the downward trend is likely. It is against this background that UK manufacturing industry must become more competitive.

1.1.2 Productivity

Productivity will be examined in relation to other G7 countries since it has been held to be a prime cause of the low competitivity of UK manufacturing and, therefore a cause of its relative decline. The 1980s was a decade when great changes were made in trade union relationships and working practices. This is reflected in considerable gains in productivity. Figure 1.6a shows the manufacturing output per head in 1970 and in 1990 at constant dollar value. Figure 1.6b presents the percentage increase in manufacturing output per employee. The percentage increase in manufacturing output per employee during the 20 years has only been bettered by France and Japan.

If this increase continues, the UK may catch up with the other countries but, although the gap has closed, the UK still remains bottom of the G7 league.

If, however, productivity is judged in output per unit of worker remuneration then Britain moves to the top of the G7 productivity league as can be seen from Figures 1.7a and 1.7b.

The high productivity per unit of pay is brought about by the low wage economy of the UK. If the UK is to be a high wage economy with an attendant high standard of living then productivity must be increased further to retain

The competitive environment

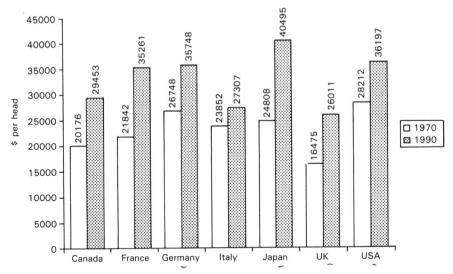

Figure 1.6a Manufacturing output per employee in the G7 countries in 1980 US dollars, 1970–1990 (*Source*: adapted from data published by the United Nations)

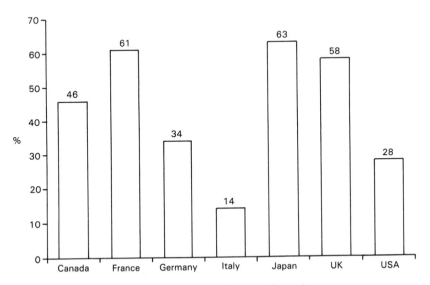

Figure 1.6b Percentage increase in manufacturing output per employee in the G7 countries, 1970–1990 (*Source*: adapted from data published by the United Nations)

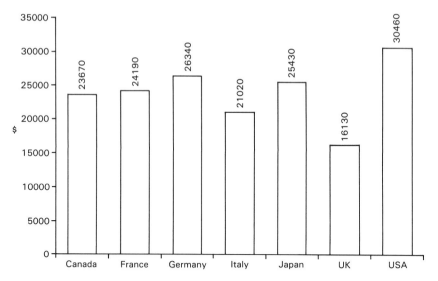

Figure 1.7a Wages per worker per year in US dollars for the G7 countries, 1986 (*Source*: adapted from data published by the United Nations)

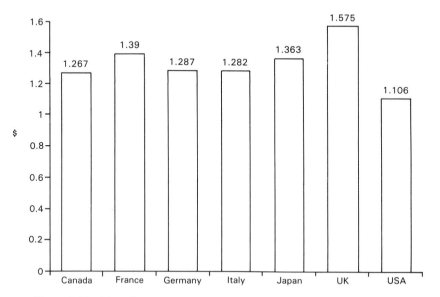

Figure 1.7b Manufacturing output in dollars per worker dollar in the G7 countries, 1986 (*Source*: adapted from data published by the United Nations)

the same cost competitivity. Increased productivity will be accompanied by a rise in unemployment unless sales of manufactured goods can rise at the same rate as the increase in productivity.

To obtain this extra sales volume British manufactured goods have to be increasingly competitive. Higher productivity itself may result in more competitive products, by lowering costs, but there is a package of competitive features which management must address.

One benefit from the low wage economy is the attractiveness of the UK to Japanese manufacturing companies as a base from which to enter the European market place. They are able to exploit the advantages of the favourable productivity per unit of wage and have resolved the package of competitive features.

■ 1.2 Global markets, global competition and global resourcing

1.2.1 The impact of global changes on manufacturing companies

This section does not set out to consider the impact of global changes on multinational manufacturing companies but to review the impact of global changes on all manufacturing companies. These changes can be expected to continue into the future.

The previous section has considered the relative importance of manufacturing in the GDP of industrialised countries. Because of developments in transportation, communication and the reduction of barriers to trade, markets are increasingly accessible. Thus, although export markets become increasingly accessible to UK manufacturers, UK manufacturers face increasing competition in what have been considered home markets.

The removal of trade barriers by instruments such as the General Agreement on Tariffs and Trade (GATT) is being extended by groupings of nations into trading areas such as the European Union (EU), and the grouping between USA, Canada and Mexico, (the North American Free Trade Area – NAFTA). Such trading areas give opportunity for companies to extend their markets. Conversely they also present the threat that other companies may expand within the enlarged market. Whether it will be an opportunity or threat will depend on gaining competitive advantage.

Changes in transportation allow rapid, low-cost movement of products and components. There are developments in all forms of transport, land, sea and air. The speed of movement will continue to increase and the cost to reduce.

Communication has undergone the greatest change with the use of electronic data interchange (EDI). Virtually instantaneous interchange of data can

Britain's Position in World Manufacturing

take place between sites anywhere in the world. As long as the communication systems are compatible, systems on one site can access systems on other sites. For example, computer aided design (CAD) facilities on one site can exchange designs with another site, where further work can be done, and the design can then be returned to the original site. EDI facilitates parent companies in an industrialised country setting up manufacturing operations in a NIC.

Changes in distribution and communication have permitted movement of raw materials, components and finished products readily and economically. Japanese companies have penetrated foreign markets with products manufactured in Japan and have only set up subsidiaries in America and Europe when it has been strategically expedient to do so. Global companies such as IBM and Philips have utilised manufacturing facilities in many countries in combination with global distribution.

Case Study Extract 1.1 shows how global influences have affected the valve spring division of the Tempered Spring Company Limited (TSCo).

Case Study Extract 1.1 The Tempered Spring Company Limited (TSCo) (Case 7)

The Tempered Spring Company Limited (TSCo) is neither large nor multinational. Nevertheless it operates globally both in sourcing raw materials and in selling its final product.

■ **Sourcing**

International sourcing of raw materials resulted when statistical process control (SPC) was applied as part of the Ford Q1.1 quality system. Variation in the quality of wire for making springs was identified as a major quality problem. Suppliers of the raw material were reluctant to develop their own processes because of the low tonnages involved, and because of the low margins imposed on the supply chain by the automotive manufacturers. As a result the material is currently sourced in Sweden, which is now the only European source of supply. Other sources of supply exist in Japan.

■ **Selling**

The automotive industry is global. In 1985 TSCo supplied to Europe and its strategy was to further penetrate the non-UK European market and to enter the US market. Both goals have been achieved.

A manufacturing facility had to be built in the US in order to supply that market. This was not because of supply difficulty from the UK but because indigenous manufacture was needed to gain acceptance.

TSCo with a turnover of £20m has become global in source of supply, in sales and in manufacturing location.

1.2.2 Competition from newly industrialised countries

Globally, NICs impact on existing industrialised countries by having low wage economies. There are differences in the wage rates quoted by different sources, but the best estimate shows a range from US$0.3 per hour in Indonesia to US$5.5 per hour in Taiwan. This compares to $10 per hour in the UK. While the UK is a low-wage economy compared with other G7 countries, it is a high-wage economy in relation to developing countries.

In addition NICs can develop their manufacturing processes at a point on the technology experience curve that has only just been reached by the manufacturing companies in industrialised nations. In developing their manufacturing processes the NICs have other advantages over the industrialised countries. There is no tradition of working practices that have to be changed. As a result the introduction of a culture that meets the needs of the manufacturing company is eased.

The workforce is hungry to raise its standard of living, which results in high productivity. Prowse in the *Financial Times*, 21 July 1993, said that $1 per hour labour in countries such as Mexico or China 'can equal or surpass American labour in simple or very complicated manufacturing tasks'.

Therefore NICs present a challenge to existing industrialised countries in manufacturing industries, by virtue of their low wage economies, and by their ability to buy 'state of the art technology'. In the case of technologically developing industries, the industrialised nations should, by virtue of a more educated population and the possession of a knowledge base, be able to retain competitive advantage in product design and technological development. However, take, for example, Thomson, which according to *The Times*, 7 November 1992, is the world's fourth largest consumer electronics group. It no longer makes televisions in Britain. It employs 24,000 Pacific Asian workers compared with 18,000 in Europe, 10,000 in North America and 10,000 in Latin America. Thomson is chasing the lowest labour costs and will switch production to lower-cost countries as skilled labour rates rise in the more developed NICs.

The Times article also questions the premise that the industrialised nations should be able to maintain a lead by virtue of possessing a more educated workforce and knowledge base. Glaxo is using Singapore as a main world production centre because it is central, pro-business, efficient and stable. It is using Singapore academics and setting up a $20 million scholarship fund. It can hire from Britain or Australia and is intent on upgrading the intellectual content of its activities in Singapore.

The transfer of the manufacturing process to NICs and developing countries may be a satisfactory solution for the directors and shareholders of the company concerned. It continues the 'hollowing out' of the UK manufacturing base and the loss of jobs and skills. The strategic question that needs to be addressed is: can management develop a managerial solution that will enable them to compete from a UK manufacturing base?

■ 1.3 Conclusion

Existing industrialised countries are witnessing a steady decline in the contribution that manufacturing makes to their gross domestic product as developing countries compete for world markets. Relatively the UK is suffering a greater decline than other industrialised countries.

Unlike some of the G7 countries, notably Japan and France, there has not been a consistent government policy in support of manufacturing industry. Currently this lack of interest in manufacturing by the government is showing some signs of change.

While productivity per worker lags behind the productivity of other G7 countries, the low wage economy of Britain means that productivity per pound of wages is high. This is one factor which has led to Britain gaining a high proportion of Japanese investment in Europe.

The increase in global competition, together with the ease of setting up manufacturing operations in Third World and newly industrialised countries, lead many manufacturing companies to a policy of overseas manufacture.

It is against this background that management faces the challenge of regenerating British industry.

2 The Manufacturing Environment

Chapter 1 has identified that British manufacturing is decreasing as a percentage of GDP faster than in other G7 countries and that Britain's share of world markets is falling more rapidly. To arrest this decline needs action by the nation as a whole but management can make a significant contribution by improving the performance of manufacturing companies. The job of management is to achieve the standard which has been termed 'world class manufacturing'.

■ 2.1 The management of successful manufacturing companies

2.1.1 Success and competition

Business success is based on developing and exploiting competitive advantage. If there is a formula for managing a company that can be used by all companies to gain competitive advantage, then only marginal advantage or disadvantage will result. As each company adopts the same approach then the only way in which advantage can be gained is by better implementation. This applies between competing companies in a single country. If global competition is considered, then it can be argued that the Japanese way of managing manufacturing is much better than the Western approach with the result that some Japanese companies have gained competitive advantage over their Western counterparts.

2.1.2 Successful companies

Peters and Waterman (1982) attempted to find the reason for success in American companies by examining excellent companies and then defining their characteristics.

To adopt either the Peters and Waterman criteria or the Japanese approach to managing manufacturing companies requires a change in the culture of the company. An example of how culture has to alter is in relation to customers. Peters and Waterman found that the excellent companies were close to customers, listened to customers and served them. Japanese companies have customer satisfaction as a main goal. British manufacturing companies are attempting to change their cultures as different methods of operating are introduced.

To gain competitive advantage Barney (1986) considered that a company's culture must possess three characteristics:

- The culture must be valuable. It must enable the company to add value, i.e. increase its store of wealth.
- The culture must be rare. It must have attributes and characteristics that are not common to the culture of large numbers of companies.
- The culture must be difficult to imitate. Other companies have to be unable to change their cultures to the required characteristics.

Thus manufacturing companies do not gain competitive advantage just by adopting the tools and techniques that are currently in vogue. Each company must identify which of the tools and techniques it should use to exploit its particular manufacturing technology and the markets it serves. The way in which the tools and techniques are to be introduced depends on the culture of the company and its ability and willingness to modify that culture. Success will result in gaining an increased customer base which can be retained in the face of competitive reaction.

It is important not to confuse company culture with national culture. Japanese companies starting up operations in the UK have not attempted to impose Japanese culture on the British workforce but, instead, to alter behaviour and habits. The Japanese idea that satisfying the customer is the most important factor in achieving competitive advantage is not new or alien to Western manufacturing. The practice seems to have been lost.

Japanese companies have exploited customer satisfaction as a strategic tool for entering and penetrating markets. Western companies can equally capitalise on customer satisfaction but to do so they need to change habits and the behaviour of personnel who have become accustomed to concentrating on other factors.

How customers are to be satisfied, and the competitive advantages to be gained are unique to each company. If all companies satisfy customers equally well then there is no advantage.

2.1.3 Lean production

Womack, Jones and Roos (1990) have coined the term 'lean production' as a description of current best practice in the automotive industry. While their research was in the automotive industry they consider the same strategy of production has been seen before in cameras, consumer electronics and motor cycles.

Consumer durables is the market sector where Japanese companies have dominated world markets and where Western companies have responded to the threat by striving to introduce aspects of the 'Toyota production system' depicted by Monden (1989a). Toyota has been described as the best manufacturing company in the world and it is the Toyota production system which has been the basis of 'lean production'.

The Toyota production system

The most important goal is cost reduction which is approached by eliminating unnecessary production activities. The goal of cost reduction is met by achieving three sub-goals:

1. Quantity control – which allows adaptation to changes in quantity and variety. This is described as just-in-time production (JIT).
2. Quality control – ensuring the supply of defect-free units from process to process. This can be termed total quality management (TQM) although many British companies take 'total' to refer to quality in every aspect of a company's operations.
3. Respect for the worker – necessary to obtain full use of human resources.

There is current emphasis by manufacturing management in the UK on JIT/TQM. The empowerment of people is regarded as an integral part of implementing these techniques but empowerment and respect for the worker are not the same thing. Are British companies intent on respect for the workers? Is the workforce a valuable asset to be trained and developed on a long-term basis or is it expendable in times of recession to be recruited again in times of boom?

The Toyota system and lean production do not rely on the introduction of new machinery, tools, techniques and methods although these may be required. They rely on a new philosophy of management that is concerned with the elimination of waste and continuous improvement in the production system.

Application of lean production

Lean production and the Toyota system as used in the automotive industry have to be tailored to be suitable in their application across the whole of manufacturing industry. For instance the application of just-in-time (JIT) as practised by Toyota and other automotive manufacturers needs a stability of demand that many manufacturing companies do not have.

15

The Manufacturing Environment

However, the elimination of waste and continuous improvement which are fundamental to JIT/TQM can be pursued by all manufacturing companies. Managements have to be selective in what tools and techniques they apply and how they change the company culture. They have to understand what can be achieved by applying JIT and how far their own company can realise those benefits. JIT has to be approached with an understanding of the limitations imposed by the process and the relationships with suppliers and customers. There also has to be a recognition that benefits will incur costs. A plan is needed to make sure that potential benefits outweigh costs. Action has to be taken to make sure that the plan becomes reality.

TQM is receiving massive publicity in the West as a way of reaching world class manufacturing status. The Malcolm Baldridge quality award in the US is seen as a prestige award, sought after by manufacturing companies, that is raising the level of quality management in companies. The European Foundation for Quality Management has been established and it presented its first European Quality Award in 1989.

Management must be clear what their company can achieve with TQM. It is not the same as product quality. It is the elimination of waste and the improvement of business performance throughout the company. A leading pharmaceutical company is aiming to achieve world class manufacturing status. One component of its world class drive is TQM but the culture change in the workforce is difficult to make because the products already achieve the very high quality standard expected from an ethical pharmaceutical manufacturing company.

Stanley Tools is proud of its quality and is working hard at achieving higher product and process quality. It is not sure that embarking on a formal TQM programme would give benefits outweighing the costs of introduction. Stanley prefers a very practical approach to quality without the label of TQM. Their approach is shown in Case Study Extract 2.1.

2.1.4 Managing the change needed to improve performance

When a company has identified what change to make to raise performance it

Case Study Extract 2.1 Stanley Tools approach to quality management (Case 6)

> BS 5750 is in place. Quality is highly regarded and, at each machine, operatives have displays of the inspection requirements to meet quality standards. But the company has not attempted to develop their acknowledged drive for quality into TQM. Their view is that the cost/benefit does not support a change in their system of achieving quality. Little use is made of SPC. Reliance for quality is based on 'ownership of the product' by the shop floor.

has to consider:

- The time scale, difficulty and cost of implementing the changes. There has to be the awareness that leading Japanese companies have been building up their achievements in manufacturing, painstakingly, over a period of 20 to 40 years. British managements have to learn from what has been done and introduce the changes in a much shorter time scale if the improvements needed to gain competitive advantage are to take place. Further, since the introduction of the changes incur significant cost, benefits are needed to justify those costs.

    However, there is a resistance to change which takes time to overcome. This time will be shortened if employees see the company threatened by competitive pressures which will lead to loss of employment. Commitment from top management to change will promote commitment at other levels and shorten the time needed for change to take place.

    There will be fears of the effects of the change. Will there be winners and losers within the workforce? If so, what will happen to the losers?

    Critical mass is needed to effect change and if resources are spread too thinly it may not be effected. Organisations need time to digest and assimilate change. This is not a plea for a long time scale but a recognition that the change process is difficult to manage. Case Study Extract 2.2 shows the resource that Kodak had to use in its successful class A MRPII implementation (see Appendix I of Case Study 4 for the definition of a class A MRPII user). Use of the 'proven path' allowed implementation to take place within a very tight schedule where completion, although six months late, was accomplished in the very satisfactory time scale of two years.
- How can the improvements be made to last after the 'driving force' that introduces the changes leaves the company? The changes will only take root if employees throughout the organisation are committed intellectually and emotionally to the changes and to developing their skills to match them.

    Perhaps even more important is how can continuous improvement become the way of life? The Japanese culture has, up until recently, provided a continuity of employment that is not present in UK, European and US companies. This continuity of employment is very important to obtain continuous improvement.

    Companies have to recognise the past history of attempts to improve management. Methods and styles of management have been adopted only to be abandoned in the face of difficulties of implementation. Future attempts have to be carried through to a

The Manufacturing Environment

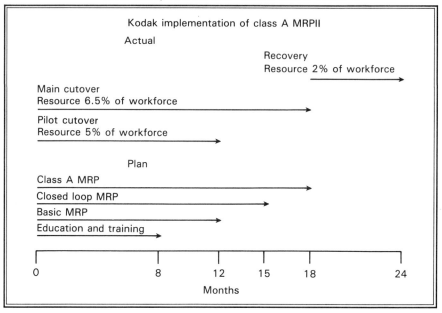

Case Study Extract 2.2 Kodak MRPII (Case 4)

successful conclusion. When new tools continue to be introduced before implementation of earlier ideas has been effective there is cause for pessimism. Doubts are being raised about TQM and Re-engineering is being advanced as the latest 'cure-all'.

- Do these changes integrate with the company's business and the results needed? Total quality cannot be sacrificed at times of high demand and shortfall in production. If the new way of managing is seen as something that can be compromised for short-term gain then allegiance to it will not be retained. The current recession has seen compromise of long-term development for short-term survival. For example, in the UK long-term training based on apprenticeships has suffered.

■ 2.2 Company experiences with the introduction of new tools and techniques

Many companies have been less than satisfied with their excursions into altering their way of manufacturing. The changes in manufacturing may have been the introduction of:

- Hardware, such as computer-controlled machines.
- Systems, such as manufacturing resources planning (MRP).

- Changes in style of management, for example, TQM involving the whole company, but making particular demands on shop-floor workers.

Such innovations involve considerable financial expenditure from which proportionately greater returns are expected. Not only have companies failed to achieve these returns but limited success or even failure has damaged the morale of company personnel. For example, there are reputedly only a handful of class A MRP installations in the UK. Of the rest, the majority fail to provide an acceptable level of control.

There is considerable evidence of companies introducing changes without considering overall strategy. This is evident on brownfield sites where change has been driven by operational considerations, rather than as a strategic response to the competitive needs of the company.

The difficulty of implementing the changes required to introduce the tools and techniques successfully can be appreciated from the view expressed by Womack, Jones and Roos (1990). They advance the theory that to be successful in implementing lean production the company must have a transplant. The reference is specifically to large mass-producers where they consider:

1. A mass-producer needs a lean competitor across the road as a demonstration of what is possible.
2. The mass-producer needs a better system of industrial finance which demands that the company does better while the considerable sums are supplied which will be needed to turn these large companies round.
3. Most mass-producers will need a crisis to prompt effective change.

There can be little doubt that a similar emphasis is needed for all companies. Case Study Extract 2.3 shows experience in the case study companies in attempts to move towards improved performance which has some of the attributes of lean production.

Success demands a strategic approach coupled with the ability of the company to change its culture so that manufacturing is managed differently. The strategic approach does not seize on the tools and techniques as a way of driving British manufacturing companies into world class manufacture. The introduction of tools and techniques appears to have been the route chosen by many British manufacturing companies, and with limited success.

2.2.1 Effective management of greenfield operations

Greenfield operations give a greater chance of success. Here management can introduce the culture that they consider gives the best chance of success. The decision taken to develop the facility will have been made from strategic rather than operational considerations. The choice of plant, of layout and methods

The Manufacturing Environment

Case Study Extract 2.3 Introduction tools and techniques

> The experience of three of the case study firms is compared:
> - Fisher (Case 2).
> - TSCo (Case 7).
> - Rolls-Royce (Case 6).
>
> Fisher faced a crisis. Circuit board manufacturing had to match the standards of other circuit board manufacturers. Failing to achieve this goal would mean outsourcing the supply of circuit boards. Accordingly, Fisher used the standards of competitors as their improvement targets. They did not call this benchmarking but it had many of the hallmarks of that approach. They did not adopt an incremental progression to improved performance which they considered could be achieved. Standards were set based on competitor performance. It was agreed that these standards had to be achieved if manufacturing was to be retained within the company. The crisis eased introduction of a new philosophy of manufacture.
>
> TSCo was faced with a crisis of declining competitivity of its product and in making a response to the large automotive companies that it supplied. The auto-line was a change in method of manufacture which had to be combined with a change in quality standards to meet customer needs and the strategy of entry into America.
>
> Rolls-Royce's advanced integrated manufacturing system (AIMS) introduction was automation of the production of aero engine turbine and compressor discs with goals set out in Section CS5.2.1 (p. 231). This was an operational decision.

of working will be designed following a strategic approach to obtaining success in the chosen market. Recruitment and training of staff can take place without a history of established work practices. The relations with trade unions can also be established in the light of strategic objectives.

Commissioning of the plant, usually, takes place before product launch and operational pressures are not high while the implementation of new and improved methods of management takes place.

2.2.2 Effective management of brownfield operations

Improving the management of brownfield operations is much more challenging. Existing operational constraints interfere with the formulation of strategic goals and with their implementation.

There will be traditional working practices and social relationships that have resulted in bad habits which need to be changed. The pattern of industrial relations will have evolved throughout the company's existence. This pattern will often be adversarial and confrontational as the unions seek to represent

Introduction of new tools and techniques

their members' interests under traditional styles of management. The changed style of management has to be sold to the unions.

Integration is needed to support new ways of working. Integration can be achieved by the introduction of changes such as:

- Strategic Business Units (SBUs).
- Team working.
- Design for manufacture.
- Emphasis on a supply chain extending from the customer, through the company, back to the suppliers.

Each of these changes provides an integration and focus on the customer. A feature common to each is a need to break down functional boundaries. For example, SBUs based on products, markets and processes weaken functional boundaries. SBUs may be divisions of companies or they may be logical groupings. Whatever form they take, the emphasis is on a common focus between marketing and manufacturing.

Team working, design for manufacture and the supply chain focus on the customer can bring about integration of company activities to meet the needs of customers.

The concept of marketing and manufacturing as the driving force of an integrated manufacturing company is shown in Figure 2.1.

2.2.3 The driving forces of an integrated manufacturing company

Marketing and manufacturing are regarded as groupings rather than as functions headed by an executive.

- The vision of the company is provided by the chief executive and the board. The vision is a framework within which marketing and manufacturing develop strategies. Marketing and manufacturing contribute to the dynamics of the vision.
- Marketing interfaces with the customer, identifies customer needs, stimulates and manages demand, manages channels of distribution and satisfies customer needs for product and service. Marketing identifies changes taking place in product technology and feeds these changes to manufacturing to make sure that core competence is developed to provide product technology.
- Manufacturing receives customer demand, whether actual or forecast and produces product to satisfy that demand. Manufacturing has the responsibility for advising on investment in core competences. It also has the responsibility for advising which core competences should be gained or abandoned through licensing, joint ventures or outsourcing agreements.

The Manufacturing Environment

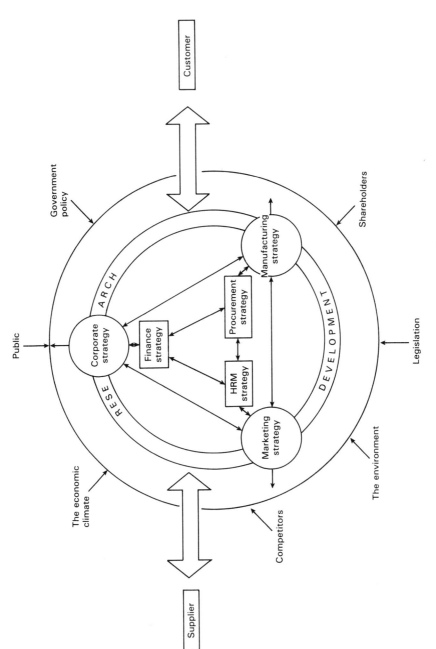

Figure 2.1 The driving forces of an integrated manufacturing company

Company environmental changes

- Research and development is directed at both products and processes. The company is regarded as being driven by corporate, marketing and manufacturing strategies. These are also the driving forces of research and development. Research and development need to be able to feed into these strategies significant technological developments which are taking place in the environment.

  Suppliers are important in the process of research and development and may possess technologies that are not available to the manufacturer. Car component design is an instance of the need for suppliers to participate in product development, for example, car manufacturers depend upon companies such as TSCo for the design and supply of valve springs.

To perform these activities effectively both marketing and manufacturing must have a strategy. These strategies will be agreed and integrated within a corporate strategy. The question then arises – what is the role of finance, purchasing and human resource management? These are strategies within the framework of corporate, marketing and manufacturing strategies. They are developed from these strategies and support them but do not drive the company.

The integrated manufacturing company is striving to react to the forces in the environment that will allow the customer to be satisfied at a profit which will satisfy the needs of:

- Employees.
- Shareholders, and other providers of capital.
- The company needs for regeneration and expansion.

■ 2.3 Changes in the environment within a manufacturing company

Changes in the technology of manufacture have the potential for transforming manufacturing. These changes can be in computer-controlled machines and systems, or in attitudes and working practices, or a combination of the two. Figure 2.2 shows the outline of how a mechanical engineering manufacturer may develop technological changes in the manufacturing process.

The development of the manufacturing process involves the integration of manufacturing support with manufacturing technology. Manufacturing technology is concerned with the operations performed on the product. Manufacturing support provides four areas of integration which overlap:

- Information integration, MRP and CAD.
- Handling integration, automated warehouses, robots and AGVs.

The Manufacturing Environment

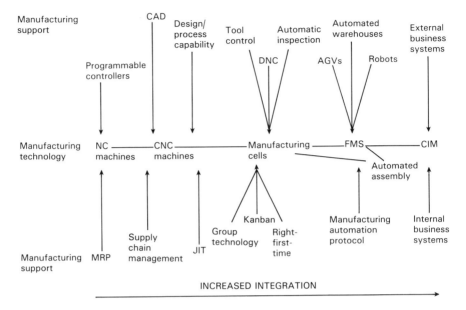

Figure 2.2 Development of the manufacturing process

- Technology integration, manufacturing automation protocol, programmable controllers DNC, tool control and automatic inspection.
- Management integration, design/process capability, supply chain management, JIT, group technology, kanban, right-first-time, internal and external business systems.

It can be seen that the development involves a mixture of hardware, systems and culture changes. This is the type of development that would take place on a brownfield site, where technology is being applied to existing processes over a period of time. The introduction of numerically controlled (NC) machines would be a first step, progressing through computer numerically controlled (CNC) machines, manufacturing cells, flexible manufacturing systems (FMS), to computer integrated manufacture (CIM). These developments require manufacturing support which may be:

- hardware, e.g. automated guided vehicles (AGVs), or
- systems, e.g. manufacturing resource planning (MRP), or
- systems combined with culture change, e.g. just-in-time (JIT).

It is not suggested that companies have to go through each stage. FMS could be introduced without any of the preceding changes, but the culture change involved would be considerably greater and therefore there would be greater risks.

The manufacturing support will not always be introduced at the stage shown but may be introduced at different phases of development, and not all may be utilised, for example, kanban may never be introduced. Nevertheless Figure 2.2 represents the complexity of changing from a traditional engineering manufacturing process to one involving advanced manufacturing technology (AMT).

Manufacturing management has not always recognised the interrelation between production hardware, systems innovation and culture change when introducing technology. For instance many manufacturing companies have spent years seeking to apply a computerised solution to systems of production and inventory planning and control. Materials requirement planning (MRP) and manufacturing resource planning (MRPII) have been introduced over many years, often with limited success. The Japanese have spent time simplifying processes, eliminating waste and applying JIT and kanban with much more success. Case Study 2, Fisher Controls Ltd and Case Study 4, Kodak Limited, describe how both have used JIT/kanban as the method of shop-floor control with the MRPII system being used strategically to relate customer demand to manufacturing capability and supplier scheduling.

Both companies simplified the flow and control of the product. While there was no attempt to schedule supplies JIT, there was rationalisation of the supplier base and an application of improved quality standards and control of incoming supplies.

Much time, money and effort have been devoted to attempts to introduce CIM. The most expensive and ambitious attempt was by General Motors in America. The introduction had seen a massive $77 billion investment in hardware and systems. 'Dozens of factories were kitted out with the best technology money could buy. Unfortunately they continued to be run in the old, inefficient manner', *The Times*, 1 March 1992.

The process of manufacture must be simplified before CIM is contemplated. The linkages between operations have to be understood both from the viewpoint of the transfer of product and the transfer of information.

Hepworth Building Products provides an example of CIM. Raw materials are limited, the process is flow line with few operations. Manufacturing processes are well developed and reliable. A major problem in automating the factory was the large bulk and weight of the product. This is complicated in the early stages of manufacture by its structural instability. Case Study Extract 2.4 shows the main activities that have to be integrated.

Although the product is simple, and the plant has few, well-developed processes, the integrating systems are complex. To apply CIM to a range of more complicated products, involving more and varied processes, leads to greatly increased complexity in the integration systems themselves.

Simplification, an integrating process itself, may make CIM unnecessary. All companies need integration of manufacturing processes and systems. How far this integration should be computerised is a judgement that has to be made by the managements of each company. Integration involves linking customer

The Manufacturing Environment

Case Study Extract 2.4   CIM at Hepworth Building Products (Case 3)

A   Systems integration − Case Study 3, Figure CS3.2

Integration is from the customer, through the process to the supply chain.

1. Sales order processing in combination with market forecasting produce a period sales forecast.
2. MRP utilises this schedule, in conjunction with stock status provided by inventory control, to produce:

   - Materials requirement schedules for purchased materials.
   - The production plan.

3. Plant operations control the prioritised dynamic schedule through the integrated process.

B   Process integration − pipe making, Case Study 3

1. Batched concrete is automatically manufactured and distributed.
2. Reinforcing cages are produced on a NC machine and fed to the pipe casting machine.
3. Glipp seals, mounted on base rings are placed into the pipe casting machine, the reinforcing cage is put on the base ring and the pipe is automatically cast.
4. The pipe is automatically removed from the machine by a robot, and moves through the process on programmed moving tables with robot removal of the base ring and transfer to testing and final curing storage.
5. The base ring is automatically cleaned, fitted with a new glipp seal and returned to storage at the start of the process.

---

demand through the production process to the suppliers. As this link occurs processes and systems are modified forcing integration to take place.

Integration of the supply chain necessitates a change towards manufacture to customer order (MCO). While this may not be possible for every company, shortening leadtimes will help a company to get nearer to real customer demand and to be more responsive to changes in volume or product mix. To support shorter leadtimes and respond to changes in customer demand requires process flexibility.

In achieving flexibility, shortening set-up time is a key area, as is the ability to manufacture the product 'right-first-time'. Right-first-time involves close co-operation between design and manufacture. The manufacturing process must be capable of making the product without subsequent rectification. At the design stage either the specification has to be within the process capability or the process capability has to be improved to meet the design specification.

From these considerations it can be seen that managing manufacturing has to be concerned not only with the introduction of new technology but also with a change in culture and systems. This change has to concentrate on five areas:

1. The supply chain.
2. Social and labour implications.
3. Introducing new technology.
4. Financial viability.
5. Cost control.

2.3.1 The supply chain

Relationships in the supply chain

The supply chain concept places the manufacturing company as a link in a chain extending from raw material supply to the final customer. The supply chain may include other manufacturers both up and downstream of the company as shown in Figure 2.3.

However, manufacturers receive supplies from multiple suppliers and in turn make deliveries to many customers, as shown in Figure 2.4.

Thus it can be seen that the supply chain concept represented in Figure 2.3 is simplistic, but becomes more complicated as multiple relations arise due to numbers of suppliers and numbers of customers. The more relationships there are then:

- The more variable the demands become.
- The more difficult control becomes.

Most companies seek to reduce the complexity of the supply chain by reducing the supplier base. This will involve some sacrifice of the 'insurance policy' of dual or multiple sourcing.

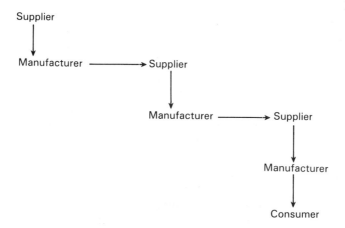

Figure 2.3 A simplistic supply chain

The Manufacturing Environment

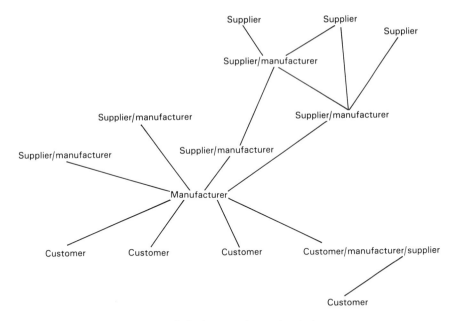

Figure 2.4 A normal supply chain

Complications in customer relationships can be eased by logical groupings of company activities into SBUs focused on product, market and technology. Simplification could be achieved by reducing the product range but this would lessen the competitive edge. Management has to look for simplification within the product range by standardising components so that customer choice is maintained but the complexity of the supply chain is diminished.

The effects of size on supply chain relationships

Within the supply chain relative size can be important. Motor vehicle manufacturers are large in relation to their suppliers. As a result they have the bargaining power to impose their demands on their suppliers. These demands will be for the supply of product to specification but they will also require suppliers to conform to their systems of quality control. The suppliers of these large companies are often not large enough in relation to their own suppliers to be able to impose the same standards further down the supply chain.

Large retailers such as Marks & Spencer can require their suppliers to meet rigid specifications of design, methods of manufacture and quality at prices determined by them. These prices are based on their knowledge of production efficiencies and profit margins in these manufacturing companies. Such large retailers are able to demand that manufacturers install systems of control on the shop floor, quality systems and EDI links providing on-line access to stock status and for product call off.

Company environmental changes

Thus, the ability to gain co-operation from suppliers can depend upon relative size. There is no doubt that the requirement of JIT supply by many large companies results in passing costs and stocks down the supply chain. Many suppliers cannot meet the JIT requirement by implementing similar JIT systems but resort to manufacture for stock and satisfying demand by 'JIT ex stock'.

Control and co-operation in the supply chain

There are advantages to be gained by moving from an adversarial to a co-operative relationship with suppliers. A first step is to reduce the supplier base to manageable proportions and then to get together with main suppliers, to explore how to co-operate in order to obtain mutual benefit.

There is a limit to the size to which the supplier base can be reduced and a limit to the ability of most companies to control other than a small amount of the supply chain, perhaps the top 10 per cent of suppliers. While the top 10 per cent may account for 70 to 80 per cent of purchases, production is still dependent on the other 90 per cent of suppliers. Figure 2.5 summarises some of the major relationships in the supply chain.

Frequently the supply chain extends from the supply of raw materials through a series of manufacturing companies to the consumer. Figure 2.5 considers the relationships that exist between a supplier and a customer. The

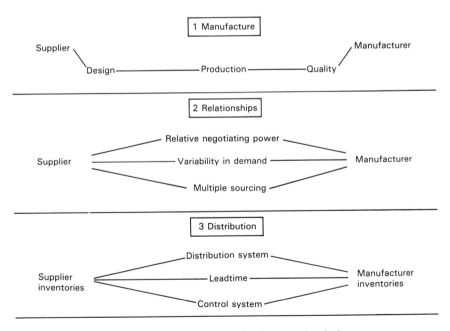

Figure 2.5 Relationships in the supply chain

customer is another manufacturer who in turn acts as a supplier. An example would be a supplier of braking systems to the automotive industry who would buy raw materials and bought-in components and, in turn supply braking systems to an automotive manufacturer who supplies cars to motorists.

Three principal areas where management needs to control the supply chain are:

1. Manufacture. The design function should extend through the supply chain with each supplier making a contribution. Quality specifications need to be agreed through the supply chain. Methods of production and systems of control to give that quality have to be negotiated between companies making up the supply chain.
2. Relationships. The relationships formed will depend on relative negotiating power. The relationship will be affected by variability of demand and decisions regarding multiple sourcing. Co-operative relationships are needed to obtain customer focus throughout the supply chain.
3. Distribution. Suppliers depend on orders placed or forward schedules to identify demand and plan their production. Real demand can be transmitted through the supply chain electronically allowing replanning of production. The introduction of EDI between companies continues to be developed.

2.3.2 Social and labour implications

The effects of technology

Companies introduce new technology to gain advantages in the manufacturing process. These advantages may be of consistency. A robot welder will always produce the same quality of weld at the same rate. The human welder will be prone to variation.

New technology may be introduced to obtain savings in direct labour. These savings will improve productivity. This can be painless if markets increase correspondingly otherwise the result is lost jobs. Ingersoll Engineers (in Mortimer 1985) foresaw the problem: 'It would be dishonest to deny that in the long term, automation in batch manufacturing will lead to a reduction in blue and white collar manpower requirements at all levels in the engineering industry'. This reduction was apparent during the eighties and accelerated during the recession at the start of the nineties.

The problem of the displacement of labour by new technology arises in all the leading economies. The Organisation for Economic Co-operation and Development (OECD) forecast unemployment of 33 million in 1993. To this total should be added the 13 million 'involuntary' part-time workers who no longer believe they will find jobs. There is a rise in long-term unemployment as a result of the displacement of semi-skilled and unskilled workers.

Company environmental changes

Gains in productivity are not the most important outcome from the introduction of new technology. While increasing productivity is a factor in gaining competitive advantage, quality improvements and flexibility must be included, as shown in Figure 2.6.

These benefits from technology strategy will only result and be sustained if the total workforce, management, support services and shop-floor workers are committed to continuous improvement. This change to an attitude of continuous improvement applies to the introduction of new technology and to the introduction of tools and techniques. Gaining BS 5750/ISO 9000 does not mean that the attitude towards quality has changed such that everyone is 'quality minded'. Frequently the objective is to obtain a mark of approval that can be used commercially in dealings with customers.

Industrial relations

The changes in working practices caused by the introduction of new technology alter accepted social structures within the company. Changes in working practices, particularly multi-skilling and flexibility, are difficult to implement in an existing plant. Industrial relations practices, built up over years to obtain advantages for sections of the workforce, become outmoded as new technology and new tools and techniques are introduced. These changes are as difficult for the various levels of management to accept as for the shop-floor workers. This may explain why the successful implementation of advanced manufacturing/CIM plants are often 'greenfield' operations.

Lucio and Weston (1992) state that the Amalgamated Engineering Union (AEU) and the Electrical Electronic Telecommunications and Plumbing Union (EETPU) which have amalgamated to form the Amalgamated Engineering and Electrical Union share the criticism of 'conflictive' industrial relations practice which pertains in most British industry. The unions accept that there is a failure of the British economy to which traditional industrial relations

Productivity improvements
Cost reductions
Quality improvements
Flexibility (ability to change what is done)
Faster throughput times

Competitive advantage

Figure 2.6 Technology strategy

Table 2.1 Survey of trade union representation in Japanese manufacturing companies in the UK

Responses from 72 companies		
No union	Single union agreement	Traditional unions
51%	37%	11%

have contributed. Workers had not been effectively involved by management in production matters. The two unions saw an opportunity in inward investment, particularly Japanese investment, to develop closer relations with management on greenfield sites.

A survey of trade union representation in Japanese manufacturing companies in the UK was carried out by the author, see Table 2.1. The vast majority of Japanese manufacturing companies setting up operations in Britain are not recognising trade unions or are entering into new patterns of relationships with them. In the survey the 11 per cent with traditional unions comprised eight companies; five were acquisitions of British companies, two were joint ventures with British companies and only one was a start-up by a Japanese company.

The relationships between management and unions and its importance in the effective management of manufacturing will be explored in Chapter 5.

2.3.3 Introducing new technology

Investment in new technology

The amount of investment by manufacturing industry in new technology is difficult to identify. Total investment in manufacturing represents both renewal of assets and investment in new technology. Figure 2.7 shows that the UK has a manufacturing investment per head 44 per cent lower than Germany and the US, 67 per cent lower than France and 170 per cent lower than Japan.

Industrial Computing (1986) gave an opinion that the UK would commit less time and money to implementing AMT strategies than its foreign counterparts. Germany was expected to make better use of AMT because its workforce was better trained and more committed.

Labour and new technology

The lower investment in technology is a cause for concern but the way in which the technology is used is an even greater cause for concern. The effective use of technology depends upon existing working practices and existing company cultures. *The Times*, 14 April 1992, speaks of the attempt at Rover to alter company culture with the offer of 'jobs for life', end of clocking and handing responsibility for hitting productivity and efficiency targets to assembly-line staff. In return the company wanted traditional demarcation lines to be

Company environmental changes

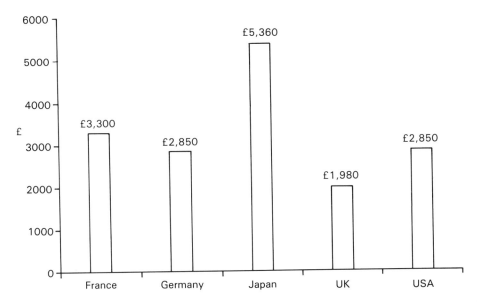

Figure 2.7 Manufacturing investment per employee (average: 1980–1990) (*Source*: adapted from data published by the OECD, Paris)

abandoned and to move the workers into teams. The deal is regarded as 'a blend of practices learnt by Rover from its Japanese partner and shareholder Honda, and modern British industrial relations thinking'.

The magnitude of the problem facing companies that have inherited traditions and working practices built up over years can be appreciated by the productivity figures quoted in the article. Nissan operating on a greenfield site at Washington was projected to produce 75 cars per worker in 1992, compared with 14 at Rover and 8 at Ford's British plants.

The Times, 12 February 1992, commented on greenfield sites: 'The Japanese have proved that with their factories on greenfield sites employing young and enthusiastic workers they can set productivity targets, and therefore profitability levels, well out of reach of established European car manufacturers.'

These statistics and comments emphasise the difficulties which face managers on brownfield sites in effecting changes which will regenerate British manufacturing industry.

2.3.4 Financial viability

The level of financial appraisal

The viability of the introduction of technology is frequently judged from an operational rather than a strategic view. Proposals for plant improvement are

The Manufacturing Environment

put forward by production managers in response to operational needs. This may be the reason for the appearance of islands of automation which many regard as being integrated at the next stage by CIM. Operationally, process improvement by the purchase of new plant or equipment is justified by operational measures such as savings in labour, rework reduction, scrap reduction, etc.

Manufacturing strategy is not well developed in most companies and piecemeal development of the manufacturing process takes place. Thus the tangible benefits of investment decided at an operational level drive investment in manufacturing rather than strategic decisions taking into consideration both tangible and intangible benefits. *The Acard Report* (1983) commented on the difficulty of quantifying intangible benefits.

The justification of capital investment

Intangible benefits may arise, for example, from increased market share that could be effected through improved design, shorter leadtime, or better quality. These benefits may be more significant than the operational cost savings that can be allocated as tangible benefits.

Simulation modelling allows companies to examine investment from a strategic rather than an operational standpoint, to consider the intangible as well as the tangible benefits and to evaluate manufacturing as a whole rather than as discrete components. Greenfield investment has to be evaluated strategically whereas brownfield investment is likely to be evaluated operationally.

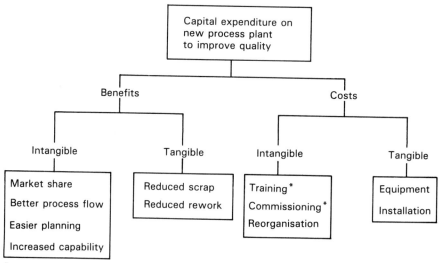

*Have been put in as intangible because of difficulties in estimating resulting, generally, in underestimate

Figure 2.8 Example of capital investment

This is another factor that may explain the high level of success of manufacturing operations developed on greenfield sites.

A problem in technology investment is the insistence by many companies on a two year pay-back for capital investment. The Cam-I survey (1988a) found that the most popular criterion for justifying investment in AMT was cost reduction, followed, in second place, by 'years pay-back'.

The assessment of capital investment is considered in Figure 2.8. The example chosen is a capital investment in new plant to give a capability which will consistently obtain the quality level required by the customer. As can be seen there are both tangible and intangible benefits arising from the investment, and tangible and intangible costs associated with the investment. The intangibles can only be based on management judgement but they must be incorporated into the model because they are usually more important than the tangibles. What-if trials can be carried out to assess the range of outcomes and to guide management on the level of risk. The appraisal does not have the apparent accuracy which quantifiable data gives but appraisal based on quantifiable data alone may be wildly inaccurate.

Pre-investment justification is required before capital expenditure is sanctioned. Post-investment evaluation is much less rigorously applied.

2.3.5 Cost control

Traditional measurement

In an era of technological change it is important that the criteria of measurement used by management should control the factors contributing to the effectiveness of the business. Further they should provide management with a firm base for making decisions as to the way in which the company should be developed.

Throughout most of the twentieth century a common method of control has been adopted by western manufacturers of discrete products. With some variation, it has been based on standard costing and budgetary control. Overhead recovery has been based on a labour hour rate.

Changes in measurement

Changes that have taken place in many manufacturing companies have reduced the importance of labour and increased the importance of machinery leading to the use of a machine hour rate. The Cam-I survey (1988a) found: 'In man-paced environments, indirect costs are most commonly recovered using standard labour hours. In machine-paced environments standard machine hours are the most popular method of recovering indirect costs. This result is not surprising. However, it is surprising that 30 per cent of companies

The Manufacturing Environment

use labour costs, in one form or another, to allocate indirect costs in their machine-paced environments'.

Companies are introducing other changes in management practice that affect the use of systems of measurement and control. When a company introduces JIT into its manufacturing it must expect to have machines idle if there is insufficient demand. It can no longer keep those machines running by making for stock. Thus a combination of short leadtimes, zero inventories and high equipment utilisation is impossible. JIT companies must be prepared to sacrifice high equipment utilisation to gain the benefits of short leadtimes and zero inventories. If departments continue to be judged on criteria of machine utilisation they automatically become inefficient in working to achieve the new goals of the company. Continuation of old systems of measurement and control will force managers back to old working practices.

Similarly with quality. It is possible with conventional systems to allocate the increased costs of introducing, for example, statistical process control (SPC) and the resultant savings in scrap and rework. But how does the system evaluate a benefit such as increased output, and the resultant saving in capital investment due to an increase in capacity? TSCo in reducing their scrap and rework by 18 per cent not only saved the direct costs attributable to the reduction, but they also increased the capacity of the line by 18 per cent.

Figure 2.9 shows the factors influencing costing systems in manufacturing companies. This figure attempts to summarise the uses that are made of costing systems and the changes that are taking place within the manufacturing environment.

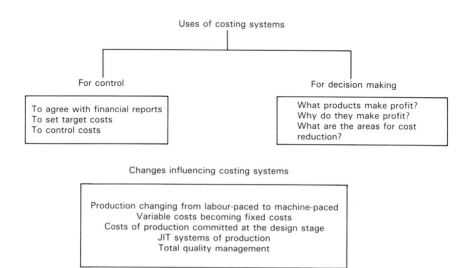

Figure 2.9 Factors influencing costing systems

2.4 How a manufacturing company competes

2.4.1 Competitive criteria

The criteria considered are those on which manufacturing has a major impact and which have a major impact on manufacturing. These criteria form a basis on which the driving forces of an integrated manufacturing company act (Section 2.3). They determine and are determined by manufacturing strategy. The core competences to be retained, developed or discarded depend on decisions made regarding these competitive factors.

Order qualifiers/order winners

Hill (1993) develops the concept of 'order qualifiers' and 'order winners'.

- Order qualifiers are attributes that the customer expects from all competing products. Some will be more critical and are termed 'order-losing sensitive qualifiers'.
- Order winners are those attributes which enable a company's products to gain advantage over those of a competitor.

Over time, order winners tend to change to order qualifiers.

An example in consumer durables is quality, which was initially an order winner. Japanese companies introduced new standards of quality and reliability that enabled them to make rapid inroads into markets for motor-cycles, hi-fi, television, cameras and cars. As other manufacturers appreciated the need to achieve the same standards of quality and reliability, which by now the consumer expected as norms, all products had to reach these standards if they were to be considered in the market place. As a result quality became an order qualifier and other factors such as price, innovation, features, service, etc. became the new order winners. Some of these have now become order qualifiers.

Hill addresses the practical definition and control of order winning and order qualifying factors and interested readers are referred to pages 44–52 of his book.

Product performance to give customer satisfaction

Kano (1993) links competitive behaviour with customer satisfaction. He identifies three levels of performance:

- Innovative product performance.
- Competitive performance.
- Basic performance.

Kano considers that there is a natural effect pulling performance down through the three levels so that innovative product performance becomes, in the course of time, the standard expected from all products. The only way to maintain product excellence is to study the customer continuously in order to provide innovative product performance.

The tool for analysing and controlling these levels of performance will be competitive benchmarking, considered in Chapter 3, Section 3.6.

Competitive factors of manufacturing

Bolwijn and Kumpe (1990) are of the opinion that, starting with the 1960s, each decade has seen a progression in the competitive factors of manufacturing.

- *Price.* The 1960s were the years of price and, therefore of cost competition. The importance attached to price and cost was reflected in manufacturing where 'efficiency management' was the goal.
- *Quality.* In the 1970s quality became the dominant competitive factor as Japanese products made increasing inroads into western markets. Quality did not replace price and cost as a competitive factor but had to be achieved in addition. Price ceased to be an order winner but remained an order qualifier.
- *Flexibility.* The 1980s saw an increase in flexibility to give a widening customer choice. The customer had a choice of product lines containing more models which were updated more frequently. At this point both quality and price became order qualifiers and a new order winner became the ability to provide the customer with a wide choice of up-to-date products.
- *Innovation.* It is suggested that the 1990s will see innovation becoming the order winner with price, quality and flexibility being order qualifiers. The competitive battle will be won by those companies that can bring technologically innovative products to the market ahead of the competition.

Figure 2.10 shows the evolution of performance criteria.

This scenario of changing competitive factors is in agreement with the analysis by De Meyer *et al.* (1987) of trends in their global manufacturing survey. Their findings are that flexibility by the end of the 1980s became the dominant competitive factor.

2.4.2 Shortening product life cycles

There is a general assumption that product life cycles are shortening and Figure 2.11 shows a typical product life cycle.

The product life cycle represents a development phase followed by launch of the product on the market, build up of volume as the market is penetrated and, ultimately, decline as the product is overtaken by a new model.

How a manufacturing company competes

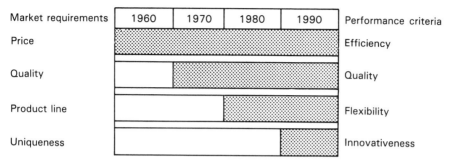

Figure 2.10 Evolution of market requirements and performance criteria for large manufacturing industry (*Source*: Bolwijn and Kumpe (1990), reprinted from *Long Range Planning*, Vol. 23, No. 4, pp. 44–57, with kind permission from Elsevier Science Ltd, The Boulevard, Langford Lane, Kidlington OX5 1GB, UK)

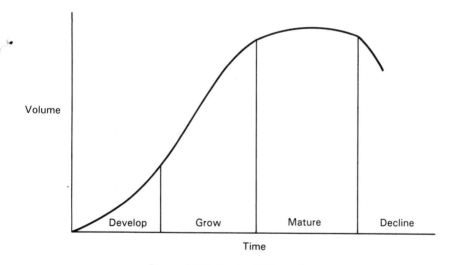

Figure 2.11 Product life cycle

Short life cycles can be seen in electronic products. Technological developments cause rapid product obsolescence as superior performance is built in to the latest product to reach the market.

Rapid changes in technology lead to shorter product life cycles in the automotive industry. Here the pressure on product life cycles comes from companies adopting strategies that they expect will give them competitive advantage. Womack, Jones and Roos (1990) identify from research in the automotive industry that Japanese models are in production for four years compared with the eight to ten years for American and European volume producers. For comparable cars the design time in Japan is 46 months and in

The Manufacturing Environment

America/Europe is 60 months. Between 1982 and 1990 the Japanese nearly doubled their product portfolio from 47 to 84 models while European volume firms reduced models on offer from 49 to 45.

Companies that supply components to car manufacturers may be affected by the short product life cycles. This will be the case for companies supplying units such as light clusters. As the model is changed, the style of the light clusters will be changed. Basic engine design changes at a slower rate and the same engine may be used in several model changes. For TSCo, Case Study 7, valve spring development is more evolutionary and new designs will be introduced when major engine modifications take place.

Some industries do not have shortening product life cycles. Hepworth Building Products, Case Study 3, have the confidence that, having identified the latest production technology and integrated the production process, the investment in a CIM plant will have a long cycle of producing an unchanged product.

For those companies that are affected by shortening product life cycles and an increase in product portfolio, then flexibility and innovation, as identified by De Meyer *et al.* (1987) and Bolwijn and Kumpe (1990), will be the strategic battleground for the 1990s. For this reason flexibility and innovation appear to be the major competitive weapon in the consumer durables sector of the market.

If companies are to compete effectively by shortening product life cycles in order to offer customers the latest technological developments, then design assumes considerable importance. Not only must design satisfy customer requirements it must also satisfy manufacturing requirements.

Tanaka (1989) considers that 80 to 90 per cent of the life cycle cost of a new product is committed at the design phase. This is shown in Figure 2.12.

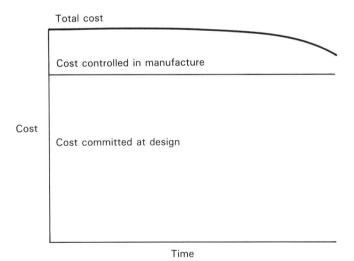

Figure 2.12 Product cost in the life cycle

Smith and Reinertsen (1991) quote a British Aerospace study reporting that '85 per cent of a product's manufacturing cost will depend on choices made in the early stages of design' and 'Rolls-Royce investigated two thousand components and found that 80 per cent of their production cost was attributable to design decisions'.

Figure 2.13 shows the allocation of life cycle costs. Eighty per cent of costs are allocated when preliminary design is available and 90 per cent when manufacturing plans are available.

The 10–20 per cent of cost that has not been committed at the design stage and which can be controlled in manufacture is likely to become more important in reducing product cost towards the end of the product life cycle. It will be at this period that problems of manufacturability will have been eliminated and where learning curve effects will be at their maximum.

However, if the life cycle is short, design has to be carried out rapidly and the product must be manufacturable immediately. Short product life cycles do not allow the product or process to be developed after launch. Learning curve effects will either not arise because manufacture will be machine controlled from the start or, if there is significant operator involvement, production runs will be so short that learning curve effects will be minimal. Costs have to be

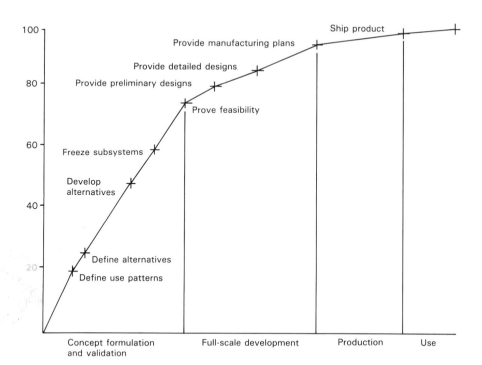

Figure 2.13 Cumulative percent of life-cycle cost determined (*Source: Developing products in half the time*, P.G. Smith and D.G. Reinertsen, Van Nostrand & Reinhold, New York, 1991, from Nevins *et al. Concurrent Design of Products and Processes*, McGraw-Hill, 1989.)

The Manufacturing Environment

controlled during manufacture. Cost reduction will be aimed at target costs for the replacement products.

Shorter product life cycles present a challenge to manufacturing management to devise a manufacturing strategy that is able to take care of the constant introduction of new products.

Care needs to be taken in considering this progression to be an industry-wide phenomenon since the model is based on large manufacturing industry. The Cam-I survey (1988a) found that 41 per cent of companies had an average product life cycle in excess of 10 years which suggests that these companies do not need to make rapid response either to competitive pressures or customer demand, but do have considerable opportunity for cost reduction during the life cycle of the product. However, there is a history of companies and industries that have appeared to be protected from market forces suddenly being faced with extinction, e.g. the Swiss watch industry with the introduction by the Japanese of cheap, highly accurate quartz watches.

2.4.3 Competing on price/cost

The efficient company has undergone changes in its approach to efficiency. The efficient company of the 1960s expected its efficiency to be based on economies of scale. The aim was for long production runs allowing modifications to be made to products after manufacture had commenced. Long production runs, in a labour-intensive environment, gave a target of steady cost reduction due to experience curve benefits. In the durables sector product life cycles were long and managements planned to prolong the life cycle further by introducing model modifications as shown in Figure 2.14.

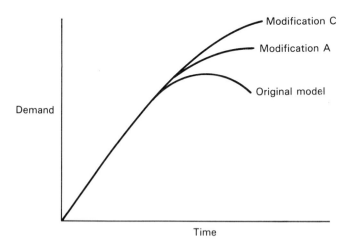

Figure 2.14 Expanded product life cycle

The situation has changed. The price at which a product will enter the market reflects a judgement of the saleability of the product in relation to those of the competition. Therefore, price reflects the premium that can be enjoyed or the discount that must be offered in order to achieve the required volume of sales. This price constrains the product cost. The product cost must be low enough to give a profit that allows the company to remain in business.

As has been stated, between 80 and 90 per cent of the cost of the new product is committed at the design stage. Thus cost and, therefore price/profit, are dependent on getting the design right. The link between design and manufacture must be right so that the anticipated costs are the actual costs.

This assumes that the costing system employed is capable of identifying product costs sufficiently accurately to compare them with anticipated costs. The Cam-I survey (1988a) found that 90 per cent of companies consider that their cost accounting practices gave a product cost accuracy of + or − 10 per cent and only 54 per cent of companies had an accuracy of + or − 5 per cent.

These levels of accuracy do not seem to be those needed to monitor actual against design costs in an era of tight profit margins.

2.4.4 Competing on quality

Many regard Total Quality Management (TQM) as the factor which will lead to the regeneration of British manufacturing. This proposition suffers from two defects.

- The first defect is that TQM is not approached by all companies as a goal which is more important than any other. The attitude that is needed can be demonstrated by considering autonomation. Autonomation means that each machine has an automatic stopping device attached which is activated when a defective piece is produced. Manual production lines have a stop button which the worker uses to bring the line to a standstill if defective products are made. Management needs to be willing to adopt such radical measures and to make the attendant effort to ensure that the same defect cannot arise again.

    Such an approach requires management to regard quality as the supreme goal. If a product of defective quality is made every other operational need must be subordinated to the correction of the process to ensure that future products will be of perfect quality. Such an approach would demand a dramatic change in the culture of many manufacturing companies.

    As has already been said, some companies are interested in obtaining quality standards approval to become an approved supplier. Others meet the quality requirements imposed on them by their main and, frequently, larger customers. In each case quality must improve but the company does not achieve a quality culture and is far from practising TQM.

TQM should be introduced because the company regards quality as an overriding commercial imperative. The company should be convinced that quality will generate customers and, if properly managed, will save money.

- The second defect is that success in manufacturing competitively has to be approached on a broader basis than TQM. It can be argued that it is impossible for companies to attack manufacturing problems on all fronts and that concentration on the most important improvement area is the way forward. When success in this area has been achieved then the next most important area should be addressed, etc. Time is too short to take a step-by-step approach of this nature. It is incumbent on management to set the climate and monitor achievement. It cannot be disputed that TQM should be a main thrust but it will not solve all problems. To maintain competitivity management must be prepared to address other issues at the same time.

The strategic focus of the company on its customers requires a company-wide approach to quality. Integration of the company's activities so that customers' needs are known throughout the company will ensure that quality is seen by everyone as their responsibility. While the achievement of quality is an operational task, the integration of company activities to obtain that quality is a strategic activity, as shown in Figure 2.1.

2.4.5 Competing on flexibility

De Meyer *et al.* (1987), from analysis of data collected in the manufacturing futures survey, consider that, while Western manufacturers were fighting the quality battle, the Japanese had already moved on to the flexibility battle. The achievement in the quality battle by European manufacturers is highlighted by Eltis and Fraser (1992): 'A recent Boston Consultancy report for the European Commission shows that in the European market 2.8 per cent of Japanese manufactured cars break down each year in comparison with 4.8 per cent of German manufactured cars, 5.6 per cent of United Kingdom manufactured cars, 7.2 per cent of French cars, and 10.2 per cent of Italian cars. There is a corresponding advantage in the quality of car components. Japanese hi-fidelity equipment, television sets and video recorders all show a continuing trend of improving quality and reliability and they have become the world market leaders'. European manufacturers have a long way to go in getting quality right before they can attack flexibility with confidence.

Flexibility can be achieved by new product introduction supported by rapid and efficient design. It can also be obtained from a manufacturing ability which can give customer choice without incurring cost penalties. The latter gives rise to the concept of economies of scope as an alternative to the economies of scale which have dominated management thinking in manufacturing. Panzar and Willig (1981) define economies of scope: 'There are economies of scope where it is less costly to combine two or more product lines in one firm than to produce them separately'. Goldhar and Jelinek (1983) identify the capabilities needed to achieve economies of scope:

- Extreme flexibility.
- Rapid response.
- Greater control.
- Reduced waste.
- Greater predictability.
- Faster throughput.
- Distributed processing capability.

Economies of scope arise in the Toyota system of production levelling where the finished car assembly line makes each type of car in succession in the smallest possible lot size.

The best Japanese manufacturers have the manufacturing flexibility that enables them to give a wider product range at competitive prices.

Flexibility is often characterised by the introduction of FMS but surveys suggest that FMS is almost a misnomer with most systems producing a limited range of products. Within that range flexibility is indeed possible. However, these are not the final products but the components of those products. New designs requiring components outside the capacity of the FMS cannot be produced. This either limits potential designs or requires expensive replacement of capital equipment.

Puttick (1987) holds the view that flexibility is needed to deal with uncertainty without limiting customer requirements. Six types of uncertainty are identified:

1. Volume variation.
2. Product mix variation.
3. Customising – even in a standard range this takes place for important customers.
4. Life cycles. Short life cycles produce uncertainty.
5. Material supplies.
6. Process reliability both of machines and people.

Slack (1991) looks at a framework for flexibility based on resource flexibility:

- flexible technology in machines and processes;
- flexible labour by the interchange of skills or interchange between processes (see Stanley Tools, Case Study 6, p. 252, CS6.5.3 Incentive schemes);
- flexibility by suppliers;
- flexibility by controllers;

and manufacturing system flexibility:

- product flexibility; the introduction of new products or modification of existing ones;
- mix flexibility; alternative product routings and fast changeover;
- volume flexibility; this is very difficult to achieve except within a limited range;
- delivery flexibility, based on a changing system of customer priorities.

The concept of time fences, Kodak, Case Study 4, is a valuable contribution to the practical application of delivery flexibility.

If manufacturing flexibility is to be achieved more is needed than just flexibility in the manufacturing process. Flexibility must be obtained in the supply chain as well. In manufacturing, where on average, more than 50 per cent of the product is bought in, a manufacturer's ability to respond to changes in customer demand is highly dependent on the ability to obtain flexibility in the supply chain. This raises questions of systems of control and relationships with suppliers. Kanban in conjunction with JIT offers a method of obtaining flexibility where there is a high level of schedule stability. Fisher Controls, Case Study 2, and Kodak, Case Study 4, give examples of flexibility using kanban on the shop floor. The support of suppliers in achieving this flexibility can be obtained but the management of the process is difficult and demanding involving the reduction of the supplier base and setting up co-operative relationships with suppliers.

How many companies are in a position to obtain supplier co-operation to make this a realistic relationship and mode of control between companies? How well can such relationships survive economic downturns and upturns? Flexibility inside and outside the company can be achieved and sustained when there is economic stability in the environment. Maintaining flexibility to respond to customer needs becomes difficult in acute recession or boom.

2.4.6 Competing on innovation

Accepting the proposition that the next move will be from flexibility to innovation, then the focus will be on research and development. The record of the Western manufacturing industries is not good in relation to its spend on R&D. The UK's record of conversion of R&D into successful exploitation of new products is particularly poor.

In some industries, e.g. aerospace, the cost of technological development is now greater than can be met by individual companies. One solution may be co-operative R&D. In some cases the company is assisted by indirect government aid, i.e. the support of large defence contracts funding some of the cost of developing commercial products. In other cases R&D is supported by direct government subsidies.

The Airbus is an example of co-operation between countries and companies. A range of planes has been developed which, according to the *New York Times*, October 1991, puts Airbus in second place to Boeing and ahead of McDonnell Douglas. The *Sunday Times*, 9 January 1994, sees the 25-year supremacy of Boeing's 747 challenged by the A340.

Innovation may be dependent upon the infrastructure, not only of one country but of several. For instance, the introduction of higher quality television pictures cannot be brought about by the manufacturers of TVs until countries agree on transmission systems.

In developing an innovative approach companies are dependent upon the strategic management of technology. The technology concerned is both product technology and process technology. Both are difficult to manage and are interrelated. In hindsight it is possible to learn lessons from unsuccessful or marginally successful developments. Some, like the Advanced Passenger Train, are being exploited by others. Supersonic aircraft and hovercraft continue but are not the success stories expected by their producers. Innovation in computing has earned considerable reward but the industry is going through a phase where future innovation is uncertain and leading companies in the industry are rethinking their strategy.

Innovation strategy can be aimed at the big breakthrough with its attendant higher risks and rewards on shorter time scales, or the incremental approach with lower risk and reward on longer time scales. In the current era of rapid technological development, management must have a policy which it is actively implementing.

2.4.7 Competing on service

Service commences from the view that 'the customer is King'. The satisfaction of customer needs for delivery and quality is paramount.

Most manufactured products require service after they have been sold. The service may be when installation takes place in order to enable the purchasing company to obtain the best benefits from the use of the product. The service may be maintenance service during the life of the product. Frequently it is a combination of the two.

Maintenance service for industrial products is important and is generally provided adequately by manufacturing companies. Maintenance of consumer products is an area that is more difficult to control. The car industry has depended upon agents but Womack, Jones and Roos (1990) suggest that there will be a move towards a closer involvement between the manufacturers and the distributive/maintenance chain to make more of service as an order winner. Evidently such re-organisation can be copied and the order winning attribute will rapidly become an order qualifier.

Voss (1992) considers service in respect to order winning/qualifying criteria and concludes that service is a qualifier, i.e. poor service has the potential of losing orders. However, the author considers that it can be an 'order winner'.

For example, new technological developments being introduced into manufacturing industry, whether they are systems, such as MRP, or hardware, such as FMS, depend for their success on close co-operation between manufacturer and customer. The resulting level of service can be sufficiently superior to that of competitors to gain advantage and thus become an order winner. As with the other ways of competing, if service starts as an order winner it will tend to become an order qualifier. Companies only retain order winners by continual improvement. Service is seen as coming from three areas:

1. The distribution chain.
2. Field (after sales) service operations.
3. The factory. Traditionally delivery performance but a wider role can be envisaged for the factory as showroom, laboratory and consultant.

■ 2.5 Conclusion

The challenge for manufacturing management is to be better than the competition in supporting products in the market place. British manufacturing companies need to sharpen their competitive edge if they are to develop new markets and defend existing markets. To be a winner against the competition a company has to be the best at meeting customer needs in order to satisfy existing customers and to obtain new customers. This does not mean that the company has to lead in everything but it must have a focus which gives it a lead in a limited range of factors which will earn a satisfactory market share.

Integration of the company requires a change in attitude and it may need a change in structure. Integration takes place when each company employee understands how they can contribute to customer satisfaction. Internally the customer is the next process. Externally it is the customer in the market place. A change in structure may be needed to allow individuals to recognise their contribution to customer satisfaction.

Products and activities have to be grouped so that there is a close relationship between customers and the unit providing the product. Strategic business units consist of such groupings aimed at satisfying specific groups of customers profitably. SBUs have to be grouped and co-ordinated so that the company obtains greater benefits than would accrue from autonomous units.

Management has to understand its business operations and identify how groups of workers can be stimulated into giving satisfaction to the customer. The principle may be the same but the practice will vary between companies in different industries, for example, a brewery, a textile factory or a manufacturer of TVs.

'In search of excellence' demonstrated principles that distinguished excellent companies. It was an analysis of the characteristics of companies judged to be excellent that produced the principles. Many of those companies are no longer excellent. The reason for their fall from excellence is more likely to result from an evolution in the principles rather than a failure of application by the companies. Management principles have been formulated and published by many writers but the search for new and improved ways of managing alters the value of some principles and makes other redundant. Continuous improvement throughout the organisation changes accepted management philosophies and methods. In relation to the theory that a manager has the ability to control only five to six subordinates, current organisational modification, such as 'delayering' levels of management, the introduction of strategic business units, together with autonomous group and team working, makes the principle meaningless.

For example, over many years, manufacturing companies have used work study to identify the 'best method' for doing the job, aware that this 'best method' will be improved by operatives as they seek to maximise earnings. Learning curve effects and wage drift are well understood in relation to shop-floor operations. Learning needs to take place at all levels of the company.

The need to gain company advantage from such improvements is vital to improving competitivity. The idea of continuous improvement in all aspects of the business is a major achievement of the managements of leading Japanese manufacturing companies. Continuous improvement is the essential goal of TQM and needs to be the goal of all manufacturing companies. However, to gain competitive advantage the company needs to improve faster than its competitors.

Gaining competitive advantage is a challenge to all levels of management in manufacturing companies. While management needs to introduce the latest technology and the most advanced systems, effective use depends on integration. Such integration depends on the organisation, its culture and the way in which its human resources are committed to a philosophy of continuous improvement.

Management needs to take a strategic view of how the company is to compete. The resulting strategy has to be dynamic in order to respond to changes brought about by competitor actions, technology advances or environmental developments. The strategy has to have the flexibility to identify and react to these changes.

Implementing the strategy may require a different attitude from both the management and workforce. That this is possible using British management and British workforces has been clearly demonstrated by successful greenfield site developments, frequently of Japanese origin.

Success has been more difficult to achieve in established manufacturing operations. Subsequent chapters will examine how managers can succeed in managing manufacturing to gain competitive advantage.

3 | Strategy, Integration and Focus

■ 3.1 The importance of strategy, integration and focus

The author considers that to obtain sustainable competitive advantage manufacturing companies must be strategic, integrated and focused on the customer. Integration is needed so that the company can react quickly to customer needs. Flexibility is necessary to react to change in the volume or mix of customer demand for current products. A strategy of flexibility requires the integration of all the activities of the company. In the past many companies were not flexible because compartmentalisation gave rise to inwardly focused functional goals. Functions tended to be closed systems with their own goals and performance measures. Open systems are required which interact to meet company goals and respond to customer needs.

Company and supply chain integration is important in the introduction of new products, in the introduction of new product technology and in the introduction of new process technology. New product design and introduction depend on companies in the supply chain keeping their processes abreast of the changes in the technology of manufacture which are taking place. They must be able to contribute to the development of new products by participating in the design of the products they supply to make maximum use of new product and new process technology. Supply chain integration is equally important to achieve customer satisfaction for the supply of existing products.

Development of corporate competence and resource is part of company strategy. Japanese manufacturing companies have effected their entrance into markets in Europe and America by integrating company activities and focusing on a specific sector of the market. The Japanese companies that have entered Western markets have concentrated their effort mainly on consumer durables sectors of the market such as motor cycles, cars, cameras, televisions, hi-fi, etc. Hence the companies are not representative of management practice

Strategy, integration and focus

in all sectors of Japanese manufacturing. Nevertheless, the philosophy and techniques of managing manufacturing within these companies have become the model which Western manufacturing companies are being encouraged to copy. This gives rise to the following problems:

- Imitation of the model, if successful, will only improve standards in Western manufacturing to those current in Japanese manufacturing. By that time the Japanese with their insistence on continuous improvement will have moved on. Even if British manufacturers succeed in catching up they do not gain competitive advantage but only parity.
- The total model may be inappropriate to companies operating in sectors of the market other than durables. Each company's management has to analyse what parts of the model are appropriate.

3.1.1 The impact of Japanese manufacturing in Britain

Between 1972 and 1992 Japanese manufactured imports into the UK quadrupled at constant prices, as shown in Figure 3.1. Over 90 per cent of these imports are in the durables sector where the Japanese have achieved significant penetration. This penetration has been with more competitive products which are the result of better management of product development and manufacture.

An analysis of Japanese companies locating manufacturing plants in the

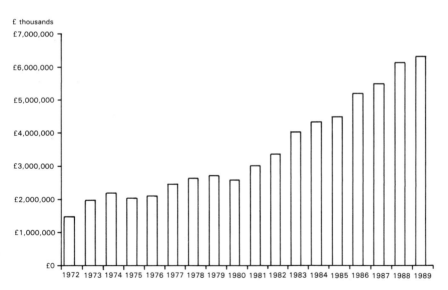

Figure 3.1 Japanese manufactured imports in constant 1987 pounds, 1972–1989 (*Source*: adapted from data published by the Department of Trade and Industry)

UK presents a similar pattern with a predominance in durables. The DTI (1991) reported that between 1951 and 1991 the UK had attracted 39 per cent of Japanese direct investment in the EU. Over 160 Japanese companies were manufacturing in Britain including every one of the ten major consumer electronic firms.

Evidently without this inward investment the decline of British manufacturing outlined in Chapter 1 would have been that much greater. Yet, if inward investment produces goods that not only replace imported Japanese manufactured goods but also contribute to exports, as is the case with Nissan, then there is a possibility that this inward investment will help the revival of British manufacturing.

The Times, 19 July 1993, states that British car component manufacturers are set to win £1 billion in new sales in Europe. The article asserts that UK car component manufacturers are the most sought-after component manufacturers in Europe. The pressure from Japanese car manufacturers setting up in the UK is given as the reason for increased efficiency and quality in the UK component manufacturers.

3.1.2 The history of the development of Western manufacturing

Adam Smith (1776) said: 'The greatest improvement in the productive power of labour, and the greater part of the skill, dexterity, and judgement with which it is anywhere directed or applied, seem to have been the effects of the division of labour'. He attributes the increase in output to:

- The increased dexterity of the workmen.
- Saving in time changing from one species of work to another.
- The invention of machinery that allows one man to do the work of many.

Smith gives an example of the benefits of the division of labour in considering pin making. Pins can be made by an individual worker or the process can be divided into operations such as drawing the wire, straightening it, cutting to length, pointing, etc. The best output for an individual worker would be 20 pins per person per day. By subdivision 4,800 pins per person per day can be produced, a 240-fold increase.

In America at the turn of the nineteenth century Taylor (1911) encouraged increasing specialisation in the metal manufacturing industries. He studied the cutting of metal and, as a result specified the feeds and speeds at which such work should be carried out. The choice of the rate at which metal should be cut was made by management not by the operative. Taylor took this principle a stage further by laying down the 'best method' of operation. This led to his proposition that 'planning' should be separated from 'doing'. The former was to be a management activity, the latter a shop-floor activity. This has been generally adopted throughout Western manufacturing industry and the operative has become responsible for carrying out the plans of others in an efficient manner.

The Gilbreths reinforced this approach by developing techniques for studying the method of working and establishing the time that the operative should take to carry out the work. Work study enabled management to move towards Taylor's concept of a 'fair day's work for a fair day's pay'. This became the basis for payment by results (PBR). Payment was either for units of product made or standard minutes produced. Since work study was not an exact science the basis of payment was negotiated between management and workers. It was in management's interest to fix the rate of payment or the standard minutes needed as low as possible and for operatives to fix the rates as high as possible.

Long production runs were welcomed by management because they minimised set-up and maximised run time leading to reduced costs and improved quality. Long production runs were welcomed by operatives because they produced favourable drift in rates due to learning curve effect.

3.1.3 Competing on the basis of low cost

The improvements in output achieved by specialisation and the attendant cost reduction was paralleled by the learning curve effect that also resulted in lowering costs. Ford developed a business strategy based on the experience curve. This was for the Model T Ford, where, between 1908 and 1926, the selling price was progressively reduced but total profits increased steadily because manufacturing costs were minimised and volume was maximised.

According to Abernathy (1978), Ford achieved the minimisation of manufacturing costs and the maximisation of sales volume by:

- Product stability.
- Capital equipment specialisation.
- Process rationalisation.
- Scale economies.
- Material specialisation.
- Labour specialisation.

However, the inflexibility of this approach by Ford meant that it took one year to change from the Model T to the Model A, during which time production was stopped.

Western manufacturing industry, generally, followed similar strategies. By the 1960s British manufacturing industry, in common with the rest of Western manufacturing industry, was competing on the basis of cost. Companies sought to rationalise product ranges in order to promote long runs and gain economies of scale. Learning/experience curve strategies were pursued to

Strategy, Integration and Focus

bring costs down. As a result the production function was under close control by cost accountants and efficiency experts who were seeking ways of reducing cost.

3.1.4 The effects of the drive for efficiency

The competitive stance of low cost production came at the end of over half a century where management had been developing ways of increasing efficiency and workers and unions had been devising ways of resisting such increases where they felt them to be in conflict with the interests of their members.

Consequently, efficiency led to a confrontational, if not adversarial, stance between management, the workforce and the organisers of the workforce, i.e. the trade unions. Shop stewards became powerful figures who had considerable influence in determining levels of manning and methods of working. In larger manufacturing companies the management, at junior and middle management level, were managers in one role supporting company goals and members of a trade union in another role, pursuing goals of self-interest. Supervisors and foremen were normally members of trade unions and were generally recruited by promotion from the shop floor. These factors led to conflict between company objectives and workforce objectives.

3.1.5 Functional organisation

The specialisation that was successful in raising the productivity of shop-floor operatives was applied to the structure of the manufacturing company resulting in a functional organisation of a form shown in Figure 3.2.

In large companies large functions arose which operated in compartments formed by their own specialism. For example, members of the quality function were trained in quality. The function set quality standards and ensured that the company's products met these standards. The quality function may have had an objective of making sure that quality standards were achieved in the most cost effective way but the overall aim of the function was to set and maintain standards of quality. Ultimately their goal was to make sure that the customer received products of the specified quality. The relationships between

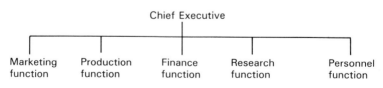

Figure 3.2 Functional organisation

Strategy, integration and focus

Figure 3.3 Compartmentalisation in manufacturing

these functional departments could be viewed as a series of compartments as shown in Figure 3.3.

Each function aimed to fulfil its own objectives. Each function had a task to perform which was done with limited communication with other functions. Activity in the functions tended to be sequential in the development and introduction of new products.

Performance measures became compartmentalised and reflected each function's measures of efficiency, rather than measures supporting company achievement. Company achievement was measured by month-end financial figures. At the end of the year these figures would undergo further modification as they were consolidated into audited company accounts.

Because manufacturing was driven by 'efficiency' and specialism was regarded as a major contributor to efficiency, functionalisation was the normal form of organisation in manufacturing companies.

The pursuit of functional goals obscured customer satisfaction as the goal of the business. The concept of trade-offs became accepted where, e.g. quality and cost were seen as trade-offs. If quality was increased then cost would increase. Therefore, with price competition being dominant and low cost needed, there was a limit to the quality that the customer could expect at that price. Skinner (1974) took the view that a factory could not perform well on every yardstick. 'These measures of manufacturing performance necessitate trade-offs – certain tasks must be compromised to meet others'. 'Such trade-offs as costs versus quality or short delivery cycles versus low inventory investment are fairly obvious'.

The concept of focus originated by Skinner continues to be one of the pillars supporting a manufacturing company in its search for advantage. However, developments in management thinking and technology make an automatic acceptance of trade-offs suspect. The trade-off between cost and quality is no longer accepted. Improved quality can reduce cost. Crosby (1989) considers that quality is free, it is only rectifying defects which costs money.

During the period that Japanese companies were gaining a significant share of the UK market for consumer durables, British companies were market led. Marketing identified the customer. The design specification was prepared and developed on marketing's perception of customer needs.

When the design was completed production was left with the task of making the product. This non-integrated approach often meant that a prolonged part of the product life cycle was devoted to taking the design that

Strategy, Integration and Focus

satisfied the customer/marketing specification and engineering it into a specification which could be manufactured.

3.1.6 The effects of lack of integration

The result of failing to get the product into the market place on time has been examined by McKinsey and Co. in Bentley (1991) who estimate that greater profit reduction is caused by late introduction of the product than by cost overrun on manufacturing or increased development costs. The importance of entry into the market on time has also been examined by Carter and Stilwell Barker (1991) who find that for a product with an expected life cycle of two years, late entry by 21 per cent of the life cycle time would result in a 54 per cent loss in revenue. While market entry may be delayed by design being late it may also be delayed by difficulties in manufacturing the design. To avoid such delays integration between functions is needed which has led to the rise of design for manufacture and concurrent/simultaneous engineering (SE). Rolls-Royce, Case Study 5, displays the benefits which can be obtained by adopting SE. The need for integration is emphasised: 'SE requires high levels of communication between participating groups. This leads to the formation of integrated teams and, if high levels of interaction are required, collocation'.

Companies cannot afford delayed market introduction while either the product or the process is modified. Where companies are faced with short product life cycles, market introduction on time of a manufacturable product becomes vital.

The problem of integrating activities within a British manufacturing company is made more difficult because of the historical influences already outlined. These influences have become embedded in a traditional manufacturing culture built up over a period of decades. Thus, in much of British manufacturing there is a strong culture which resists integration.

3.1.7 Japanese manufacturing companies

In the West the Toyota, or lean production, system is taken to be the model of effective manufacturing. This is the model which has given Japanese manufacturing companies competitive advantage over Western manufacturing companies.

It comes as something of a surprise when reading the books of Shingo (1983, 1986a, 1988) to find that the thinking of Taylor and the Gilbreths has been underpinning the Toyota production philosophy. The fundamental difference between the American (Western) and the Japanese (Toyota) approach is in the consideration of the aims of Taylor and the Gilbreths. In Shingo's view the West has confused the means with the ends. Means were thought of as ends and the quantitative aspects of work study have been used

Strategy, integration and focus

for setting standards for payment rather than work study being used as a tool for improvement.

Shingo (1988) views Taylor's work as fundamental improvement and waste elimination and that Western management has lost sight of this goal of doing away with waste. Concentration on economic lots has obscured the possibility of reduction in set-up times. Opportunities for basic improvements are overlooked.

Shingo criticises the confusion that he considers exists in the West between process and operation. His view is that the difference is:

- Process is the flow of products between workers.
- Operation is where work is done transforming the product.

The process/operations matrix is shown in Figure 3.4. Processes link operations. Process is the priority area for improvement where non-value added activities such as transportation, waiting time and delays can be eliminated. Concentration can then be on operations. This is the basis of the Toyota production system.

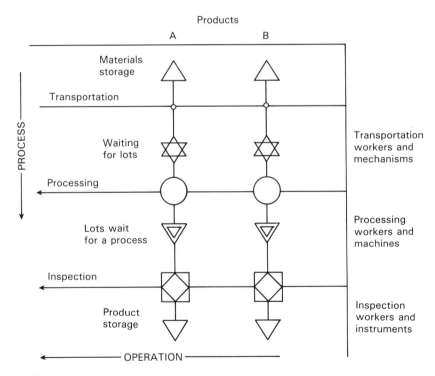

Figure 3.4 Processes–operations matrix (*Source*: adapted from *Non-stock Production* by Shigeo Shingo. English translation copyright © 1988 by Productivity Press Inc., PO Box 13390, Portland, Oregon, 503-235-0600. Reprinted by permission)

Strategy, Integration and Focus

■ 3.2 The strategy of an integrated manufacturing company

The focus of an integrated manufacturing company is shown in Figure 2.1 and the importance of manufacturing strategy in obtaining this focus will be examined.

Corporate, marketing and manufacturing are the strategies that drive the manufacturing company in relation to markets and in relation to changes in the environment. Manufacturing strategy is considered, not production strategy. Production is taken to be the narrow range of activities concerned with making the product. Manufacturing is considered to be all the activities from receipt of customer demand to satisfaction of that demand as shown in Figure 3.5.

The distinction between manufacturing and production is important in considering how to develop and implement systems of management and control that will promote effective co-ordination. It is not suggested that the area identified as manufacturing should come under the control of a single manager. Nevertheless, companies must be able to co-ordinate the activity of manufacturing with the activities of marketing and distribution. This co-ordination has to allow the company to reconcile its objectives of profit and market standing with the objective of customer satisfaction.

The triangle of strategies shown in Figure 3.6 does not mean that there is a fixed relationship. Corporate strategy must integrate marketing and manufacturing strategies. The integration of the strategies focuses the company on its markets. Corporate strategy does not dictate the other strategies. It may at times formulate a framework within which marketing and manufacturing strategies are developed but at other times it will respond to strategies made in those areas. Similarly marketing and manufacturing are not in a fixed relationship to each other. Marketing may take a lead when a market opportunity is identified but manufacturing may take a lead when technological developments of either product or process can provide a competitive advantage.

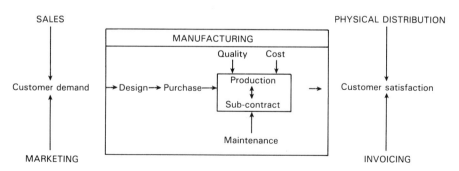

Figure 3.5 The activities of manufacturing

3.3 Developing strategy in an integrated manufacturing company

3.3.1 Strategy and operational pressures

The company has to maintain a strategic approach in the face of operational pressures. Internally these operational pressures result from the necessity to meet budgets. The planning and control systems in most companies are driven by short-term profit and profitability. Externally, for public companies, additional pressures are imposed by the stock exchange, the analysts and the investors who are not looking at strategy that may produce jam tomorrow but at the operational results that will produce jam today.

For many companies an annual review of the company's goals takes place. Such a review is frequently projected forward over a time horizon, typically three to five years. Integrated with this strategic appraisal is the annual budget where the strategic plan is updated annually and the first year becomes the budget. Problems which arise are that formal strategic thinking only takes place annually and the budget may rapidly become out of date. Rhefeld (1990) suggested that six-monthly budgeting had considerable advantages. 'To me, six-month budgeting was just twice as much work. But after three or four fiscal periods, I began to appreciate it. I came to welcome the opportunity to change the budget because the world had changed so much in six months'. Rhefeld was in the personal computer business. Not all businesses face such rapid change. Nevertheless, focusing on the reasons why the budget is becoming outdated raises questions beyond operating performance.

3.3.2 Developing the strategic framework

In the manufacturing company reaction to strategic change is complicated by the very large investment in plant, equipment and people skills which makes rapid change difficult. Figure 3.6 demonstrates the way in which manufacturing strategy needs to provide a framework that integrates and develops:

- Product design.
- Process design and development.
- Supply chain maintenance.

The resources in these areas prescribe what products the company can manufacture and what markets it can serve. The way in which the infrastructure of manufacturing operates will determine the reaction time of the company. This reaction will be to changes in the environment such as actions by competitors, changing customer needs, alterations in legislation and the influence of pressure groups.

Strategy, Integration and Focus

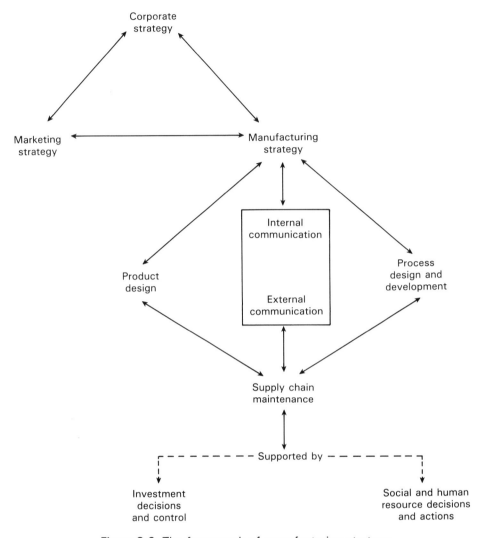

Figure 3.6 The framework of manufacturing strategy

To be successful in co-ordinating these areas the company must have effective internal and external systems of communication:

- Electronic systems such as CAD–CAM provide links between product design and the process of manufacture. Manufacturing planning and control systems integrate design, supplies and the logistics of production. EDI links demand and supply through the supply chain.
- Organisational communication systems can be improved by restructuring the company.
- Involvement of people. This may result from restructuring, e.g. the

Developing strategy in an integrated company

introduction of cell manufacture. It may also result from cultural changes in the company, e.g. brought about by TQM.

Manufacturing strategy is supported by:

- Investment decision and control. Manufacturing is the major user of finance in the company and the products it makes generate the revenue. A problem is that developing the long-term manufacturing strategy that makes use of long-term funding is aimed at producing long-term revenue. This may adversely affect profitability in the short term.
- Social and human resource decisions and actions constrain manufacturing strategy decisions and in turn are affected by such decisions. For example, the drive by UK manufacturers to match productivity levels of other countries is a necessity if manufacturing is to be competitive. It is, however, a contributor to the steep increase in unemployment that took place in the recession at the end of the 1980s and the beginning of the 1990s.

Restructuring manufacturing involves HRM decisions, e.g. changes in skills and work roles may result.

3.3.3 Formulating the strategy

In making an annual review there are approaches that focus on the needs of the business and approaches that concentrate on the environment in which the company operates:

- Gap analysis is an example of focus on the needs of the business. It is a consideration of forward profitability and is usually based on Return on Capital (ROC). A datum of required ROC is compared with that which the business will produce if it continues into the future with its current activities unchanged (see Figure 3.7). A gap arises between the profitability needed by the company and the projected profitability. This gap can be closed by considering alternative strategies. The search for the alternative strategies should not be, but often is, dominated by quantitative analysis.
- Environmental analysis concentrates on the business in its competitive environment. A commonly used model for this purpose is Porter (1980), who considers that there are five forces driving industry competition: rivalry among existing firms; bargaining power of suppliers; bargaining power of buyers; threat of new entrants; threat of substitute products. Since the model concentrates on the competitive forces within the industry it is likely to develop a broad strategic approach that will need subsequent quantitative analysis.

Strategy, Integration and Focus

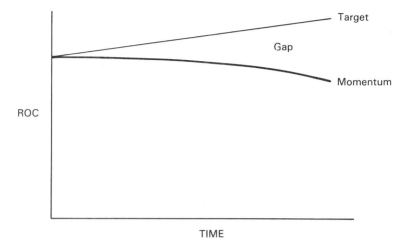

Figure 3.7 Gap analysis

These approaches will be supported by a strengths, weaknesses, opportunities and threats (SWOT) analysis. SWOT analysis will be of products and markets, as shown in Figure 3.8 and will give management an insight into:

- The status of existing products in existing markets.
- New products that are needed to support existing markets.
- New markets that can be exploited using existing products.
- New products that can develop new markets. This will result in diversification.

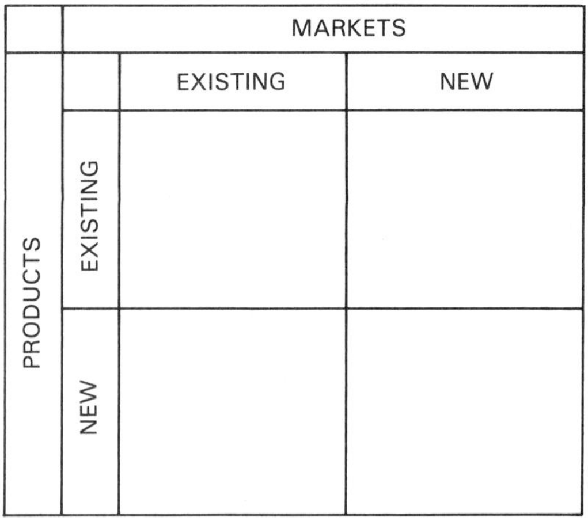

		MARKETS	
		EXISTING	NEW
PRODUCTS	EXISTING		
	NEW		

Figure 3.8 Analysis of products and markets

3.3.4 Strategy formulation and implementation

Where in the organisation strategy should be formulated and implemented is important both from the standpoint of integrating activities and of gaining commitment to the plan. Commitment is a vital factor in gaining successful implementation. Many excellent strategies have been formulated but implementation has either been imperfect or, at the stage of implementation, the plan has been rejected totally. Many theorists have proposed as a solution wide involvement of company personnel in both strategy formulation and implementation, advocating that the process should be a 'bottom-up' rather than a 'top-down' activity. In large organisations there are major difficulties in adopting this thinking unless the organisation is restructured.

The structure of the organisation

The multi-product company, whether national or multi-national is increasingly structured on the basis of SBUs. Illustration 3.1 demonstrates the way in which Asea Brown Boveri, a huge enterprise, was organised into small units with profit and loss responsibility and meaningful autonomy.

The restructuring of organisations into autonomous units is frequently accompanied by a reduction in the number of levels in the hierarchy to produce a flatter organisation, Illustration 3.2 considers the application in General Electric. Flattening the organisation reduces layers of management which improves communication. The flatter organisation compensates for the reduction in levels of management by promoting team working and people empowerment involving a delegation of decision making.

Whether the company is large and global or is small and local there is much to be gained by organising into SBUs. There is a strong motivational influence where SBUs formulate and implement the strategy for marketing and manufacturing their products. Both GE and ABB decentralised their organisations. GE retains large groupings which have been made leaner and fitter by removing layers of management. ABB retains company structures where real profit and loss responsibilities are preserved.

Capital employed and revenue generated are easier to identify in a SBU which has unique products and markets served by unique resources. Nevertheless, strategic decisions made at SBU level have to take into account the use of shared resources, e.g. research and development, central procurement, company administration, etc. It is important that valid allocation of the costs of these shared resources is made, with which SBUs agree.

There is difficulty in carrying out industry analysis in multi-product companies that operate in more than one industry. SBUs are industry specific which eases formulation and implementation of strategy.

Strategic change is a complex process involving an assessment of the environment, the competition, the capability of the company and the politics obtaining within the company. Functional organisation promotes internal

Illustration 3.1 Organisation in ABB

Percy Barnevik explains ABB's matrix system, a structure designed to leverage core technologies and global economies of scale without eroding local market presence and responsiveness. We know what core technologies we have to master, and we draw on research from labs across Europe and the world. Being a technological leader in locomotives means being a leader in power electronics, mechanical design, even communications software.

The organising logic of ABB

ABB Asea Brown Boveri is a global organisation of staggering business diversity. Yet its organising principles are stark in their simplicity. Along one dimension, the company is a distributed global network. Executives around the world make decisions on product strategy and performance without regard for national borders. Along a second dimension, it is a collection of traditionally organised national companies, each serving its home market as effectively as possible. ABB's global matrix holds the two dimensions together.

At the top of the company sits CEO Percy Barnevik and 12 colleagues on the executive committee. Reporting to the executive committee are leaders of the 50 or so business areas (BAs), located worldwide, into which the company's products and services are divided. The BAs are grouped into 8 business segments, for which different members of the executive committee are responsible.

Functional coordination teams meet once or twice a year to exchange information on the details of implementation in production, quality, marketing and other areas. The teams include managers with functional responsibilities in all the local companies, so they come from around the world. These formal gatherings are important, but the real value comes in creating informal exchange throughout the year. The system works when the quality manager in Sweden feels compelled to telephone or fax the quality manager in Brazil with a problem or an idea.

Alongside the BA structure sits a country structure. ABB's operations in the developed countries are organised as national enterprises divided into nearly 1,200 companies with an average of 200 employees. These companies are divided into 4,500 profit centres with an average of 50 employees. Separate companies allow you to create real balance sheets with real responsibility for cash flow and dividends. With real balance sheets, managers inherit results from year to year through changes in equity. Separate companies also create more effective tools to recruit and motivate managers. People can aspire to meaningful career ladders in companies small enough to understand and be committed to.

Source: Copyright © 1991 by the President and Fellows of *Harvard Business Review*; all rights reserved. Reprinted by permission of *Harvard Business Review*. The logic of global business: an interview with ABB's Percy Barnevik by William Taylor, March–April.

Developing strategy in an integrated company

Illustration 3.2 Organisation in GE

In 1981, Welch (Chairman and CEO of General Electric) declared that the company would focus its operations on three 'strategic circles' – core manufacturing units such as lighting and locomotives, technology-intensive businesses, and services – and that each of its businesses would rank first or second in its global market.

But scale alone is not enough. You have to combine financial strength, market position, and technology leadership with an organizational focus on speed, agility, and simplicity. First we took out management layers. Layers hide weaknesses. Layers mask mediocrity.

Cutting the groups and sectors eliminated communication filters. Today there is direct communication between the CEO and the leaders of the 14 businesses.

Review systems

At our 1986 officers' meeting, which involves the top 100 or so executives at GE, we asked the 14 business leaders to present reports on the competitive dynamics in their businesses. How'd we do it? We have them each prepare one-page answers to five questions: What are your market dynamics globally today, and where are they going over the next several years? What actions have your competitors taken in the last three years to upset those global dynamics? What have you done in the last three years to affect those dynamics? What are the most dangerous things your competitors could do in the next three years to upset those dynamics? What are the most effective things you could do to bring your desired impact on those dynamics?

Five simple charts. After those initial reviews, which we update regularly, we could assume that everyone at the top knew the plays and had the same playbook. It doesn't take a genius. Fourteen businesses each with a playbook of five charts. So when Larry Bossidy is with a potential partner in Europe, or I'm with a company in the Far East, we're always there with a competitive understanding based on our playbooks. We know exactly what makes sense; we don't need a big staff to do endless analysis. That means we should be able to act with speed.

Probably the most important thing we promise our business leaders is fast action. Their job is to create and grow new global businesses.

Source: Copyright © 1989 by the President and Fellows of *Harvard Business Review*; all rights reserved. Reprinted by permission of *Harvard Business Review*. Speed, simplicity, self-confidence: an interview with Jack Welch by Noel Tichy and Ram Charan, September–October.

Strategy, Integration and Focus

politics. Re-organisation into SBUs, profit centres and teams will reduce internal politics. When they arise their solution will be made easier.

3.3.5 Strategic intent

Hamel and Prahalad (1989) use this term for the strategic outlook which will put a company into a position of competitive advantage at a point in time in the future. Strategic intent is a vision of where the company is going.

Case Study Extract 3.1 shows how the formal strategy of TSCo, an SBU, was made to reflect its strategic intent. Without a vision of the future strategic planning can become nothing more than an analysis of the company's existing business and strategy development will become merely long-range projections of this business.

Case Study Extract 3.2 sets out a vision statement for Fisher Controls. As the case study shows, this vision statement is supported by detailed strategies to which analysis, such as that outlined in Section 3.3.3, has been applied. The vision statement provides a long-term goal which is not constrained by existing products and processes.

Hamel and Prahalad (1989) consider two models of strategy:

1. Maintaining strategic fit which matches ambition to available resources. Resources are allocated to product–market units.
2. Leveraging resources to reach demanding or seemingly unattainable goals. In this case investments are made in core competences as well as in the product–market units.

Case Study Extract 3.1 TSCo implementation of the auto-line for valve spring manufacture (Case 7)

> The strategy which led to the introduction of the auto-line had a long gestation period in which top management were aware of a need to alter the direction of the business. They were conscious that this would involve change in manufacturing away from batch processing to continuous flow processing. During this period there was strong pressure from the car manufacturers for improvement in quality and reduction in price.
>
> The management also knew that to be successful in introducing the auto-line the climate for change within the company had to be favourable. The long wait before installation was spent changing the culture in the company. The decision to go ahead with the auto-line was only made when management judged that the workforce was in favour of the change.
>
> The final strategic plan which was produced was more a justification for the strategy than an analysis which brought about the strategy. The strategy had been formed as part of the company vision of their development as a leading valve spring manufacturer.

Developing strategy in an integrated company

Case Study Extract 3.2 Fisher Controls worldwide vision (Case 2)

The Fisher worldwide vision is:
- We will view the world through the eyes of our customers.
- We will be a global, market driven, quality enterprise.
- We will create a competitive advantage for our customers through applying products and services.
- We will create an environment in which employees and other stakeholders can rely on the integrity of our commitments and where all employees are given the opportunity to realise their full potential.

Stalk *et al.* (1992) argue that it is not core competences which are critical but capabilities which they define as a 'set of business processes strategically understood'. Honda's success, which Hamel and Prahalad attribute to the core competence in engines and power trains, is attributed by Stalk *et al.* to capabilities such as an expertise in 'dealer management'.

There is an argument for using the concept of adding value through the supply chain as the guide for the company mission. The value chain, as analysed by Porter (1985b), encourages management to focus on ways of increasing added value at all stages of the supply chain.

Strategic intent views the business in terms of 'where are we now?' and 'where do we want to be in the future?' Whether it is core competence, strategic capability or value added in the supply chain which are considered to be the forces which drive strategy, all may be examined by the technique of strategic benchmarking.

3.3.6 Strategic benchmarking

Hall, Johnson and Turney (1991) give a definition of competitive benchmarking, derived from Xerox as 'the continuous process of measuring our products, services, and practices against our toughest competitors or those companies renowned as the leaders'.

Comparison has been attempted in the past. The Centre for Interfirm Comparison collected data from subscribing companies under agreed conditions of confidentiality and comparability. The Centre analysed the data and reported the results anonymously to subscribing companies.

There are companies that specialise in analysing published company results to provide main performance indicators on an industry basis.

Benchmarking differs in that it is not a subscription service but is a voluntary exchange of information between companies. The ultimate benchmark is the leader. The steps are:

- Identify superior performance in critical activities, this will be from published data.

Strategy, Integration and Focus

- Expand information through direct contact and visits to selected firms by small multi-functional operating teams.

It is not sufficient to chase the leader because, as the target is achieved, it will be surpassed by a new leader. There can be a deliberate attempt to surpass the standard and take over leadership. Benchmarking is a learning process and its impact may be seen in Illustration 3.3. The statement by the CEO does not examine how but what must be done. How the goal is to be achieved is a learning process.

A similar analysis is reported by Walker (1992). The unit manufacturing cost at Xerox was equal to the Japanese US selling price. Benchmarking showed the need for an 18 per cent increase in productivity over a five-year period in order to catch up compared with a planned increase of 8 per cent.

These examples demonstrate the need for radical improvement in crisis situations.

3.3.7 Technology strategy

A strategy to enhance technology is vital to the manufacturing company. Technology is both the technology of the product and the technology of the process. Porter (1985a) suggests an approach based on his generic strategies of differentiation and overall cost leadership.

Focus utilising either differentiation or cost leadership

Product technological change and process technological change are concentrated on the market strategies developed by the company. The strategies are regarded as being exclusive and as a result need different skills, resources and organisational requirements.

Just as the Japanese have cast doubts on the validity of trade-offs in manufacturing so they have been able to pursue strategies resulting in both differentiation and cost leadership. To achieve this position in the market place they have used product technological change in conjunction with process technological change. A high profile example is the developments made in the automotive industry which have been termed 'lean production'.

Illustration 3.3 Benchmarking

> Chaparral Steel identified that Pohang, a Korean steelmaker, had become the benchmark for landed costs in the US. The CEO of Chaparral then stated 'Our objective is to get our labor costs below the cost per ton of the ocean voyage from Korea. That way, they can pay their people zero and we can still meet them at the unloading dock with a cost advantage'.

Source: Hall, R.W., Johnson, H.T. and Turney, P.B.B. (1991) *Measuring Up: Charting pathways to manufacturing excellence*, Business One Irwin, Homewood, Illinois.

Porter thinks that competitive advantage can be gained from technological developments that do not involve technological breakthrough but which are in fact quite mundane. This can be seen to apply to the developments in process technology made by the Japanese where they have gained competitive advantage from perfecting ideas that originated in Western manufacturing. Radford (1989) in describing how Sumimoto transformed Dunlop Tyres comments '... we also benefited from the Japanese skills in production efficiency and their meticulous attention to detail'. Illustration 3.4 highlights how the Japanese gained competitive advantage by the use of manufacturing technology.

Prahalad and Hamel (1990) in their examination of the core competences of manufacturing companies are highly critical of the role of SBUs in technology strategy. In their view SBUs are in the business of getting competitive products into the market place but 'no single business may feel responsible for maintaining a viable position in core products nor be able to justify the investment required to build world leadership in some core competence'. There is the problem that an SBU may develop competences that will be retained by the SBU. This arises because the loyalty of SBU members is to the unit rather than the company. Prahalad and Hamel propose the development of a strategic architecture that will force the organisation to identify and commit to technical and production linkages across SBUs to provide a distinct competitive advantage.

However, as can be seen in Illustrations 3.1 and 3.2, both ABB and GE build their technology strategies on decentralised organisation structures. They appear to have a strategic architecture in place which co-ordinates company-wide technology strategies.

Speed is a vital ingredient in business reaction and is particularly difficult to obtain in manufacturing companies with the inertia of systems, skills and

Illustration 3.4 Competitive advantage by the use of manufacturing technology

> Then Japanese semiconductor manufacturers entered the 16K dynamic random access market (DRAM) with the product based on a U.S. 4K DRAM design, the prior generation of product technology. Through superb process and materials technology, they were able to increase manufacturing yields significantly beyond industry norms and were able to offer parts at a significant price advantage and ten times the quality of U.S. sources. The value of the quality improvements was enormous to electronic manufacturers. They no longer had to sift through thousands of devices, weeding out bad ones, or repair assembled boards failing due to poor quality RAMS. As a result, the Japanese captured significant share of a heretofore U.S.A. dominated market.

Source: Quinn, J.J. (1985) How companies keep abreast of technological change, reprinted from *Long Range Planning*, Vol. 18, No. 2, pp. 69–75, with kind permission from Elsevier Science Ltd, The Boulevard, Langford Lane, Kidlington OX5 1GB, UK.

Strategy, Integration and Focus

resources. When these have been committed for one strategy there is difficulty in obtaining rapid change to respond to a new strategy. The type of organisation and control in ABB and GE can give a faster response.

Sourcing decisions define core competence

Venkatesan (1992) considers core competence from the viewpoint of sourcing decisions which, while they arise at an operational level, are strategic in nature. By outsourcing manufacturing activities core competences can be lost. Once lost they can almost never be regained.

A focus on competences can be gained by making outsourcing decisions at the level of subsystems rather than components. Posing the question what subsystems are indispensable to the company's competitive position over subsequent product generations focuses attention on how the company differentiates its product offering from those of its competitors.

Subsystems which must be manufactured within the company comprise families of components which are manufactured by product technologies of strategic importance to the company. Making such an analysis enables the company to relate its core competences to the overall strategy of developing products and markets, i.e. its strategic intent.

■ 3.4 How to achieve focus

Manufacturing strategy enables a company to focus on market opportunities. These may arise either from marketing pull or technology push. Having the right manufacturing capability for the markets that the company serves, or which it wishes to enter, is the objective of the strategy. In analysing manufacturing capability it is important that this matches the needs of the company's markets.

Since the technology of manufacture differs, e.g. between a printed circuit board manufacturer and a concrete pipe manufacturer, the application of management philosophies and systems will also differ. Puttick (1987) developed a framework for analysing the market based on an analysis of the effects of complexity and uncertainty. These are defined in Figure 3.9.

- Complexity is concerned with the number of individual items that must be sourced and controlled.
- Uncertainty outside the business arises from the uncertainty of demand for products. Inside the business it arises from the unreliable behaviour of people, machines and systems.

Using these two factors it is possible to identify four market segments that make different demands on manufacturing, see Figure 3.10.

How to achieve focus

COMPLEXITY	UNCERTAINTY
Number of items per product	Product variety
Number of levels in the bill of materials	Sales volume
Degree of commonality of parts	Quality
Number of sequential operations in a manufacturing routine	Plant availability
Number of workcentres	Absenteeism

Figure 3.9 The manufacturing dilemma (*Source*: Puttick (1987))

Companies may serve more than one segment. In the example of ABB it will be located in all four segments of the market analysed. Manufacturing strategies need to be developed by the subsidiary companies, the SBUs or profit centres to focus on the individual markets.

The importance of competitive criteria will not be the same in each of the four market segments identified in Figure 3.10. Puttick used the competitive criteria of:

- Fitness for purpose.
- Service.
- Price.

His assessment of their importance in each of the four sectors is shown in Figure 3.11. To achieve products that meet these competitive criteria requires a strategy that focuses on the segment of the market in which the product will

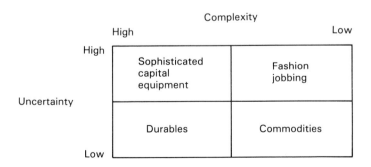

Figure 3.10 Product–market focus (*Source*: Puttick (1987))

71

Strategy, Integration and Focus

	Complexity	
	High	Low
Uncertainty High	Fitness for: purpose 1 service 2 price 3	Fitness for: purpose 1 = service 1 = price 3 =
Uncertainty Low	Fitness for: purpose 1 = service 1 = price 1 =	Fitness for: purpose 3 service 2 price 1

Figure 3.11 Competitive stance (*Source*: Puttick (1987))

be offered. A manufacturing capability has to be developed which will be able to satisfy the competitive stance better than the competition.

The case studies identify the differences in competitive criteria between companies. Case Study Extract 3.3 examines the competitive criteria for Hepworth's entry into the concrete drainage pipe market compared with Fisher Controls' criteria for printed circuit board manufacture.

All manufacturing companies can better assess the factors which will gain competitive advantage by being close to their customers. Shortening manufacturing and supply chain leadtime will get companies closer to the customer which will get them closer to current demand and reduce the factor of uncertainty.

Case Study Extract 3.3 Competitive criteria

- ***Hepworth Building Products***

  The CIM plant, built to manufacture concrete drainage pipes incorporated state-of-the-art technology and latest product design. The capability of the plant matches the product design. The plant was developed anticipating a low rate of change in the products. Although the products lie within the commodity sector, product differentiation will give better 'fitness for purpose' but the plant will also produce at lower price.

- ***Fisher Controls***

  The printed circuit boards are to be incorporated into process controllers where there is a continuous upgrade in functionality. The customers, other manufacturers, are looking to obtain competitive advantage by installing these controllers. Rapid introduction of new designs is important and problems of manufacturability, in many cases, have to be overcome after design.

3.5 Integrating the manufacturing company

Integration is a key capability to enable a manufacturing facility to respond quickly to customer needs. Figure 3.12 presents a model of the integration of a manufacturing company. The model and subsequent discussion suffer from the problem that company integration cannot be dissected, neatly, into discrete areas. For example process integration and integration horizontally have considerable overlap. With these reservations the model is a powerful tool for stimulating managers to examine the integration of their own company.

3.5.1 Process integration

The following areas are all to be considered in examining the integration of the process:

- Product flow.
- Manufacturing capability.
- Rapid changeover.
- Physical connection of operations.

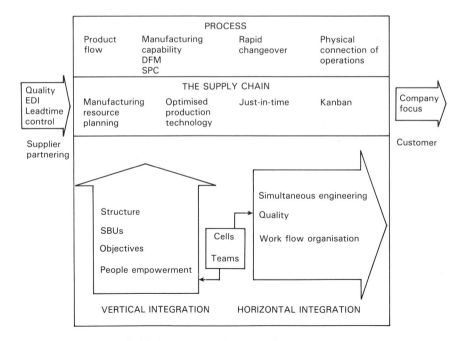

Figure 3.12 Integrating the manufacturing company

Strategy, Integration and Focus

Product flow

Work flow organisation is the basis for process integration but can only be successfully applied if product design and manufacturing capability are matched to give right-first-time quality. This needs design for manufacture (DFM) in conjunction with the development of statistical process control to enhance manufacturing capability.

In combination with product layout the development of cells integrates the manufacturing process to respond to the customer.

Figure 2.2 shows the development of the integration of a mechanical engineering process. It can be seen that manufacturing process development depends on the development of systems of manufacturing support.

Simplification is necessary for integration. Simplification arises from designing a smooth flow of product from the incoming supply chain through the manufacturing processes to the customer. In the course of designing this work flow integration takes place. The integration feeds back into the MPC system which, as it produces more accurate output, assists process integration.

Case Study Extracts 3.4, 3.5, 3.6 and 3.7 give different examples of simplification and integration.

These case study extracts give examples of how integration takes place. They show simplification of the process not of the operation. The operations remain technologically complex. This simplification of the process substantiates Shingo's view that non-value added processes should be eliminated allowing concentration on operations.

Case Study Extract 3.4   Hepworth Building Products (Case 3)

> The operations of manufacture are carried out on proven equipment. Each piece of equipment carries out complex operations. This equipment is laid out in a simplified process which is computer integrated. The CIM strategy outlined in Case Study 3 allows customer demand, the manufacturing process and material supplies to be integrated as shown in Figure CS3.2.

Case Study Extract 3.5   Fisher Controls (Case 2)

> The MPC system was MRPII which was applied to shop-floor control while production was in a process layout. As a result of implementing the manufacturing strategy the production layout was changed to product.
>
> The simplification of work flow and the development of teams emowered to control quality resulted in JIT production. Control of production devolved to the shop floor and the execution module of the MRP system became redundant.

Integrating the manufacturing company

Case Study Extract 3.6 Kodak (Case 4)

> The MPC system which was in operation at Kodak was to be superseded by a MRPII system. The intention was to gain class A user status within eighteen months. Shop-floor control was to be by kanbans.
>
> Implementation of the MRP system was complex and demanded considerable resource to achieve the goals set. Control of shop-floor operations with kanban greatly simplified the implementation. The end result was a system being used strategically for linking supply chain activities.

Case Study Extract 3.7 Rolls-Royce (Case 5)

> The advanced integrated manufacturing system (AIMS) was based on a number of four-axis, CNC turning centres. These complex machines reduced the number of operations from 21 to 5 and the leadtime from 22.5 to 8 weeks. The machine tool population was halved (62 to 31) and the scrap rate was reduced by 40 per cent.
>
> AIMS was deployed into 27 cells connected together with handling equipment to give automation of work flow. The simplified process could now be computer controlled to form a computer integrated system.

Simplification of the processes leads to an understanding of how the operations can be connected into a work flow.

In connecting operations it is important that each product leaving a process should be perfect otherwise the JIT shop-floor control exercised by Fisher Controls, Case Study Extract 3.4, and Kodak, Case Study Extract 3.6, will become ineffective. In the same way the coupling of customer demand to the supply chain at Hepworth, Case Study Extract 3.5, involves plant failure if defective product is made. Rolls-Royce, Case Study Extract 3.7, producing in batches of one, is dependent on that batch being perfect. Thus it can be seen that the achievement of perfect quality is a necessary part of work flow organisation leading to integration.

Simple operations which are operator controlled need the error-proofing type of control instanced in Illustration 4.5 in the production of hi-lift springs at TSCo. More complex operations which are machine controlled need inbuilt systems of error proofing for the production of perfect quality.

Manufacturing capability

The main contribution made by improving manufacturing capability is the ability to manufacture right-first-time. This results in a reduction in scrap and rework. As a consequence, work flow is simplified and the manufacturing planning and control system is better able to control shop-floor operations.

The effects of simplification are cumulative. As the improved manufacturing capability reduces the amount of scrap and rework so the schedules become more accurate and customer delivery promise can be fulfilled. As a result customers have confidence in delivery promises and cease to build insurances into their demand. Finally material requirements planning is simplified.

Rapid changeover

Conditions are now right for an attack on set-up time with a view to single minute set-up. It is important to be clear as to what is meant by single minute set-up. It is the time that the machine ceases to produce, not the time involved in set-up. Shortening set-up is accomplished, partly, by setting-up off the machine. For example DSF, Case Study 1, are looking for the reduction in set-up time by improving activity in the die preparation area, by standardising press beds so that all machines accept all dies and by moving towards a cassette die so that adjustment after change is minimal.

Set-up can also be shortened by introducing NC machines. CNC machines improve the speed and quality of set-up even more. Direct numerical control (DNC) of machines allows designs, carried out on CAD, to be directly loaded onto machines. As can be seen these stages in development shorten set-up and also integrate activities.

Physical connection of operations

Cells will aid transfer of products between operations in the cell since the cell layout is based on work flow. The transfer of product between cells can be integrated by the use of FMS or CIM plant.

Rolls-Royce, Case Study 5, and Hepworth, Case Study 3, present examples of FMS and CIM. In the case of Rolls-Royce, it is an integrated cell. In the case of Hepworth, an integrated factory. Technologically manufacture is totally different but the operations are automatically connected together. Rolls-Royce uses AGVs and pallets. Hepworth uses automatic handling robots and moving tables.

TSCo, Case Study 7, developed an auto-line which contained the same operations as its batch line. Some of these operations were improved to give individual treatment of valve springs. Great benefit was obtained by the integration of the process achieved by connecting the individual operations with handling equipment. The result was automatic transfer of product from one operation to the next.

It can be seen that in each of these examples the non-value adding processes have been the main area for gains.

3.5.2 Integrating the supply chain

The historical approach to supply chain management

Manufacturing companies for years have attempted to integrate the supply chain by some system of manufacturing planning and control (MPC). Such systems have aimed to relate customer demand to manufacturing capacity and to material and component procurement. Prior to the computer these systems depended upon planning boards, forecasting techniques, economic batch rules, etc. backed by human judgement based on experience. Typically, weekly meetings were held to review backlogs and priorities, leading to updated plans. These meetings issued revised production schedules.

Priorities were established for production and purchasing. Progress chasers were employed to realise these new priorities and revised schedules. Management looked for solutions to the problems of production and inventory planning and control by simplification and rationalisation of the product range to improve the repeatability of manufacture. Set-up frequency was reduced and long runs gave scale economies. Standardisation of the product and its components reduced the range of materials and components to be purchased or made. With a more repeatable standard range of products production processes could be standardised.

Rationalisation led to conflict with the marketing function which considered that the ability of the company to compete in the market place was eroded by the restricted choice presented to customers.

Stock holding was utilised to decouple demand from supply. Finished goods stock was intended to meet unexpected increases in customer demand. Raw materials and component stocks were intended to protect the company against increases in demand as production had to be increased to replace the depleted finished goods stock.

Raw material and component stocks were also a protection against the uncertainty of delivery as well as a protection against the delivery of substandard, unusable supplies.

Evidently decrease in demand would lead to rises in stockholding. The increase in stock could be made worse due to the adversarial relationships that existed between manufacturers and their suppliers. Cancellation of orders was difficult and unwanted supplies of previously ordered materials may have had to be accepted.

Rules were built up to aid planning such as economic order quantities (EOQs). These were used to relate the quantity to be ordered, the cost of placing the order and the cost of stock holding.

Similarly economic batch quantities (EBQs) for manufacture related the cost of set up, the batch quantity and the cost of stockholding.

Manufacturing a component, which involved operations on different machines with different set-up times, would give different EBQs. The size of batch, therefore, became a compromise.

Forecast of demand took place at different levels. Forecasts would be prepared for future supply of products. These forecasts were uncertain because of the length of time they had to be projected into the future to take account of the long leadtimes of procurement and manufacture. As a result the historical usage of materials and components would be analysed and used as the basis for ordering most supplies.

The stocks held were in three categories:

1. Raw materials and component stocks were held to guard against uncertainties such as:
   Failure by suppliers to meet delivery promise.
   Unforeseen increases in the level of production.
   Failure of purchased items to meet quality specification.
2. Work-in-progress brought about by:
   Processing material in batches.
   The need to obtain maximum machine, operator and departmental efficiency which led to queues of work in front of machines.
   Priorities that arose after the launch of the manufacturing programme. Work already started was left and became WIP, while other work was launched and rushed through the process.
3. Finished goods stock was accumulated:
   To have goods available to meet uncertain customer demand.
   As the result of maximising machine, operator and departmental efficiencies.

Overall the manufacturing system was managed in such a way that stock was accumulated at all stages from raw materials to finished goods. The effect of this stock accumulation was that:

- Systems of control were complicated.
- Large quantities of capital were tied up in stock.
- Stock was accumulated which, because of extended leadtimes, often became obsolete as customer needs changed.

Development of manufacturing planning and control systems

The Western approach Most companies in the West addressed the problem of MPC by looking at systems inside the company. The search was for a system which could take customer demand, and transform it into a manufacturing and purchasing programme. The purchasing programme would be the basis for placing orders in the supplier network. A better MPC system would improve the purchasing programme but the relationships with suppliers would continue on the existing adversarial basis.

Customer demand could be forecast in the case of make-for-stock manufacturers. For make-to-customer-order companies the demand would be actual. Many companies had a mixture of the two systems of manufacture.

For manufacturing companies that had complex control situations it had always been recognised that the quantity of data to be processed was a problem which could only be resolved when computers became sufficiently powerful. With this development computerised MPC systems have been implemented in American and British manufacturing companies.

The Japanese approach The Japanese approached the control of production and of the purchasing programme differently:

- The development of quality control had taken them down a path of eliminating waste. This had brought about simplification of processes and systems that dramatically reduced the control problem.
- In Japanese manufacturing industry first level suppliers are an integral part of a final manufacturer's network, frequently with interlocking shareholding. This led Japanese companies into a further resolution of the manufacturing planning and control problem. They developed integrated control from suppliers through the manufacturing process to the customer.

These approaches led to a development of JIT and kanban systems.

Current use of MPC systems

A brief review will be made of systems for MPC.

Network analysis Network analysis is important for the control of projects. In manufacturing it is mainly used in the segment of the market identified by Puttick (1987) as sophisticated capital equipment. The products will have a large number of sub-parts and manufacturing will take months rather than weeks.

The network of activities is drawn to determine those that can be carried out simultaneously and those where the preceding activity has to be completed before commencement of the succeeding activity. The network will be analysed in a computer programme to determine the critical path, i.e. the chain of activities that prescribes the length of time for completion of the project. Management can make decisions on shortening the critical path by, for example, increasing resource allocation.

Network analysis can be used for control as well as planning.

MRP/MRPII Materials requirement planning (MRP) developed into manufacturing resource planning (MRPII). The original MRP system was solely an ordering tool that rapidly evolved into the closed loop MRP shown in Figure 3.13. The basis of the closed loop system is a feedback from monitoring the

Strategy, Integration and Focus

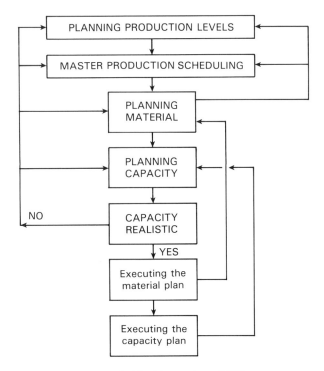

Figure 3.13 Closed loop MRP

outcome of the plan which causes replanning to take place and is based on the concept of a control cycle shown in Figure 3.14.

Before attempting to purchase and install a MRP system managers should analyse what they want to control. Is the requirement sufficiently understood to define what input a computer system needs, and what output is expected? Is the input data available and is the integrity of the data adequate for the system? When these questions can be answered confidently, in the affirmative, the time is then right to buy and install a package. Even so satisfactory installation and commissioning will require significant resource and commitment.

The closed loop MRP system has been further developed by incorporating a 'strategic' element. This is the system termed MRPII. In the case of Fisher Controls, Figure CS2.5 of Case Study 2 presents the MRPII system which was developed. Fisher Controls became able to control its shop-floor production using JIT not the MRPII programme and as a result the company has concentrated on enhancing the strategic use of the system. This has enabled the company to integrate the international scope of its business in order to improve its response to customer demand.

Many companies have not been able to implement MRP systems satisfactorily. Sometimes the failure has been due to attempting to computerise existing manual systems. These did not work as manual systems and failed to work

Integrating the manufacturing company

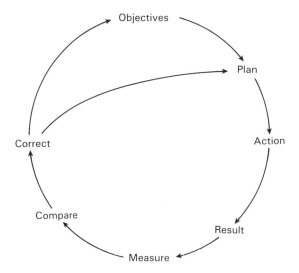

Figure 3.14 The control cycle

as computerised systems. Sometimes this failure was because they were unable to achieve the levels of data integrity needed to make the MRPII system work. Yet more failures arose because MRPII was looked on merely as a system. Lack of commitment to making the system work arose because people preferred the existing familiar system, whatever its imperfections.

Optimised Production Technology (OPT) OPT is a proprietary software package that allows a company to simulate its production system. Details of the manufacturing facilities need to be fed into the package. The experience of one company was that this was a major task since plant records did not represent the facilities accurately nor the output that could be achieved from those facilities. There was also a problem of data integrity.

The problems of data integrity are comparable between MRP systems and OPT. Use of the two systems, however, is different. OPT uses finite capacity planning. OPT simulates the production system and uses the demand made by the product requirements to produce a schedule. This schedule highlights the bottlenecks. The advantage claimed is that the 'simulation' can be re-run using changed information to obtain the best match of production requirements against plant capacity.

The bottlenecks are the areas where management need to concentrate their action.

The philosophy of OPT is very specific in its use of three measures:

1. Throughput. The rate at which money is generated by the system through sales.

Strategy, Integration and Focus

    2. Inventory. All the money that a firm invests in purchasing things which it intends to sell.
    3. Operating expense. All the money spent in order to turn inventory into throughput.

These measures can be improved by the steps shown in Figure 3.15.

- Throughput can be raised by improvements at bottlenecks. Improvements at non-bottlenecks will not increase throughput.
- Material flow improvements will reduce inventory.
- Effecting those improvements *and* adhering to the schedule will reduce operating expense.

It can be seen that the philosophy of OPT is very similar to the waste reduction outlook of Japanese manufacturing but with the discipline of a computer system. Supporters of OPT advocate a change to throughput accounting which is reviewed in Chapter 6.

Just-in-time production and kanban control   The JIT production system originated in Toyota. The production system was developed over a period of 25 years. This is very pertinent for Western manufacturers to consider when attempting to develop JIT introduction over a much shorter time scale. As can be seen from Kodak, Case Study 4, and Fisher Controls, Case Study 2, JIT has been developed to a level where it can be used to control shop-floor operations.

OPT RULES
1. Balance flow not capacity
2. The level of utilisation of a non-bottleneck is not determined by its own potential but some other constraint in the system
3. Utilisation and activation of a resource are not synonymous
4. An hour lost at a bottleneck is an hour lost for the whole system
5. An hour saved at a non-bottleneck is just a mirage
6. Bottlenecks govern both throughput and inventories

Bottleneck improvements	Material flow improvements
• Offload the bottleneck	Attempt to improve material flow in all areas of the plant
• Do not process scrap	Remove disruption by reducing:
• Do not produce scrap	• set-up times
• Drive out idle time	• set-up variability
• Reduce the process time	• scrap
• Reduce the set-up time	• machine downtime
MAXIMISE THROUGHPUT OF A BOTTLENECK	**ALWAYS ADHERE TO THE SCHEDULE**

Figure 3.15 The philosophy of OPT (*Source*: Courtaulds Filament Yarns)

JIT started as a manufacturing system that produces the units needed, at the time needed and in the quantity needed. JIT has to be controlled through the various stages of manufacture so that succeeding operations are fed JIT by preceding operations. A form of flow line is needed so that raw material moves through the stages of manufacture to become a finished product JIT to satisfy the customer. The control of manufacture is by a kanban system.

Kanban is the Japanese for 'card' or 'signal'. Kanbans are often cards that authorise withdrawal of a unit batch from storage at the end of one stage of manufacture and place it in storage in front of the next stage of manufacture, ready to be processed. Kanbans are used to authorise the previous stage of manufacture to make the next unit batch. It can be seen that for JIT and its control system to work there needs to be a balance of manufacturing time between the stages of manufacture.

If the process is to respond to the mix variety in customer demand on a JIT basis then it must be able to manufacture in that mix variety. Changeover from one product to another must be rapid and all processes in the flow line must be capable of achieving similar rates of changeover. JIT production requires JIT supply. Supply and production JIT can only be obtained if there is relative stability of demand in the short term, i.e. a month, typically.

Controlling the incoming supply chain

MPC systems convert the independent demand of the customer into a dependent demand of materials and components. These will either be purchased outside the company or manufactured within.

Thus the MPC system produces a schedule which has to be satisfied by supplies obtained from the supply chain. Changes are taking place in the attitude of companies to supply chain relationships. These changes are from an adversarial to a co-operative or partner relationship. The adversarial relationship was based on searching for the best value for money at the time the order was to be placed. Companies customarily sourced supplies from a number of suppliers as an insurance policy against interruption of supply. The suppliers, and potential suppliers, who satisfied the company's criteria of suitability would have orders placed on the basis of best value for money.

The relationships of the Japanese manufacturing companies with their suppliers have influenced many British manufacturers to change the way in which they manage the supply chain. Simplification is needed if the supply chain is to be effectively co-ordinated and controlled. A reduction in the supplier base has been a first move in the process of simplification.

The reduction of the supplier base should be in conjunction with use of a Pareto analysis to identify the 20 per cent of suppliers who provide 80 per cent of the company's supplies. These are the suppliers where there is value to be gained in moving towards a relationship where the supplier can be regarded as an extension of the company.

Effectively this change in relationship is from purchasing goods from a

Strategy, Integration and Focus

supplier to buying manufacturing capacity. Thus, changing the demands made on that capacity do not affect the supplier as long as the changed programme can be accomplished and the same profit earned from use of that capacity. At this stage supply chain integration starts to become vertical integration. This form of vertical integration does not need the ownership of the supplier. Rolls-Royce use the term 'revenue sharing partnerships' to describe this form of integration in the supply chain.

The formation of these types of relationships should lead to benefits. For example a more stable relationship should mean that the supplier can achieve shorter and more reliable leadtimes. Shorter leadtimes from suppliers in combination with shorter leadtimes through the process will lead to better control of the whole supply chain.

There are, however, doubts about the strength of these relationships. Large manufacturers are able to enter into and sustain co-operative relationships with their, generally, smaller suppliers. A report in *The Times*, 21 December 1993, stated that a letter had been sent by Ford to its suppliers demanding price cuts and carrying the threat that failure to meet this demand may result in the supplier not being asked to bid in the future. Smaller manufacturers face the difficulty in the demands being made on them by big customers and their inability to transmit these demands to their own large suppliers.

The trust and co-operation that are the basis for these new supplier relationships do not appear to be strong in British manufacturing industry. There are examples where supplying companies have worked at building up relationships with a customer. They have developed a satisfactory product but, before they have the opportunity to reach an acceptable price, the business has been transferred elsewhere. Subsequently when this price from the new supplier rises the business returns. It is not being argued that customers should be tied to suppliers if there are cheaper sources of comparable supplies available but that an existing supplier should be given the opportunity to lower their costs and prices.

Improving control of the supply chain

Bringing quality under control Manufacturing schedules that are issued with allowance for the production of excess units to take care of possible scrap or rework can result in finished product either being over- or undermade. In the case of overmake unwanted product has to be stocked which may or may not be of use. In the case of undermake measures have to be taken to satisfy the customer order. If right-first-time quality can be achieved manufacturing schedules can be issued for the exact quantities needed.

The same thinking can be applied to purchase schedules. On the other hand if the quality of incoming goods is variable then the MPC system has to work in conjunction with inflated safety stocks. The planning and control system becomes unnecessarily complicated as it has to reschedule materials

Integrating the manufacturing company

and, possibly, manufacture. Uncertainty affects plans. Processing more information does not help. What is needed is to get quality right and get stability in the MPC system.

Bringing leadtimes under control Leadtimes both internally and through the supply chain have to be shortened and made consistent. Long leadtimes can give rise to artificial orders as customers attempt to get in the manufacturer's queue. The manufacturer will adopt the same policy with its suppliers. Long leadtimes give an extended forecasting horizon that will make for a less accurate forecast.

Consistent leadtimes eliminate a further control problem. Reliable leadtimes allow the planning and control system to concentrate on variability of demand. If leadtimes are uncertain the system has to deal with this additional variable.

Improving communication in the supply chain The planning and control system can be improved if there is better communication through the supply chain. Better communication will arise if there is co-operation between supplier and customer. Demand through the supply chain will vary depending on the demand for the final product. It is this demand that has to be made visible through the supply chain.

Take the case of a manufacturer of castings supplying a customer making engines who in turn supplies the manufacturer of tractors who then sells to farmers. This supply chain is shown in Figure 3.16.

The manufacturer needs knowledge of the customer's expectation of forward demand. The supplier of castings needs knowledge of the manufacturer's schedules not only to programme its manufacturing but also to manage its own supply chain. Trust must be established if this openness is to exist. The way in

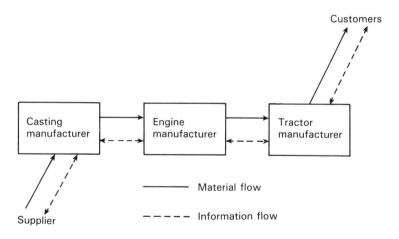

Figure 3.16 Material and information flow in a supply chain

Strategy, Integration and Focus

which information is communicated between companies in the supply chain is moving to electronic data interchange (EDI). It is interesting to speculate whether these close relationships in the supply chain represent management of the supply chain or a move to a form of vertical integration which supplements the example given on p. 84.

Improving value added in the supply chain As discussed, one way of obtaining competitive advantage is to eliminate non-value adding activities in the supply chain. A potentially more advantageous way to gain competitive advantage is to examine where value adding activities can be improved or can be introduced. Benchmarking is a potent tool for carrying out this examination. Advantage may arise by looking outside the company's industry and examining practices which are adopted in other industries. For instance the way in which distribution is effected by other industries can give a greater improvement in a company's distribution system than examining competitors' practice.

In attempting to catch up with the best of the competition a company will improve its competitive stance. Benchmarking gives potential for original, entrepreneurial thinking that can give advantage even over the best of the competition. Such original thinking is open to all companies.

3.5.3 Integration vertically within the company

As discussed on p. 63 organisations are being restructured into autonomous units with a flatter organisation which reduces layers of management giving improved communications. Within this form of organisation the aim is to involve all company employees in the strategy of the company and focus on the needs of customers and the needs of the company. Satisfying these demands is a prime activity of the cells and teams. Delegated decision making, referred to as 'people empowerment' will happen at team and cell levels.

The flatter organisation is simpler and people can understand how it works. However, there is a downside to flattening the organisation. Management jobs become redundant and future promotion opportunities are reduced.

In the larger, more complex company, particularly the global organisation there is a need for a further structure to give organisational integration. This is provided by a matrix organisation which integrates the profit centres and SBUs. This matrix is of great importance if development taking place in one part of the organisation is to be transferred effectively throughout the company.

Within organisations there are certain functions that will be retained centrally, such as purchasing, where the expertise provided by centralised purchasing together with a concentration of buying power can give the company competitive advantage. The personnel function will be retained centrally as a policy-making activity that will set the company style and standards. At operational level the management of personnel is undertaken by supervisors and managers.

Objectives

Figure 3.14 presents the control cycle as a closed loop system for controlling objectives. This control cycle provides a basis for individuals to control their own objectives. At the company level it provides the basis for controlling individual objectives that, in sum, make up the company objective.

The company can succeed in meeting its objective only when subordinate objectives are agreed at all levels and across the organisation. The organisational forms outlined above allow objectives to be agreed, integrated and controlled. The people undertaking the activities can be empowered to make decisions in their sphere of operations. With ownership of the activity goes responsibility which is the accountability for success or failure. There is a move towards rewarding success by profit-related pay schemes. Current statistics show a considerable increase in the number of these schemes.

Since company strategy is formulated and implemented in a dynamic environment, there has to be revision of objectives as changes take place. It is necessary to implement these changes rapidly if they are to be effective. Flat organisations with rapid communication and decentralised responsibilities can give this response.

Objective setting Systems of objective setting have been tried before. Management by objectives (MBO) was widely introduced into American and British industry, commerce and the public sector during the 1960s and 1970s. MBO did not meet with long-term success in most organisations. Some companies had two or three attempts at making the technique work but, in most cases, MBO was either abandoned or continued in a considerably modified form. Some of the terms can be recognised in companies that are attempting to develop the tools and techniques currently in vogue. Key result areas were a feature of MBO and are again in favour. TQM advocates continuous improvement that was the basis of improving business results in MBO.

The author's critique of what went wrong with MBO may be salutary for managers attempting to introduce some of the new tools and techniques currently on offer:

- The term technique has been used deliberately because this was a major reason for the failure of MBO, and is a reason why many companies are disillusioned with their attempts at the introduction of current new methods. To be successful these new methods must be regarded as philosophies, as a way of life. Kodak, Case Study 4, demonstrates that the introduction of MRP does not succeed by approaching it as a technique. It shows the way in which company

Strategy, Integration and Focus

culture has to change, and also displays the amount of effort and determination that is needed to gain success.
- The approved way of introducing MBO was via management development advisers who were either internal or external advisers. As a result MBO was not 'owned' by the company. It was supplementary to the mainstream activity of the company and, when the going got tough, was expendable. In the same way some companies regard TQM, JIT, etc. as nice ideas that ought to be promoted so long as they do not interfere with the real task of the company which is to get product out of the factory gate.
- MBO became very mechanistic. Objectives and plans for their execution were committed to paper. Formal reviews were scheduled at fixed intervals when objectives were to be updated. People found themselves having to change objectives to meet the demands made by changes in the environment but these changes were not made through the MBO system. As a result the MBO system became out of line with reality.
- Objectives were agreed between superior and subordinate. This presented the problem that objectives are not achieved by individuals but by groups which interact across functional boundaries. For example, a key result area for a production manager could be the accomplishment of a production programme. Success depended on support by maintenance to give the required level of uptime and by purchasing to obtain the quantity of the right quality material. Objectives were set in these areas independently. Lack of co-ordination often frustrated achievement of individual objectives.

This brief, and personal, critique is aimed at managers who are thinking of introducing new methods generally, and objective setting in particular. Ownership of the new methods has to be promoted and the culture of the organisation has to be supportive of their introduction.

While Kodak, Case Study 4, used a consultant in implementing MRPII, the consultant was not seen to be responsible for the system. That responsibility was owned by the company, the team and the individuals. To gain that ownership a massive programme of education and development had to take place.

Changes leading to vertical integration

The case studies show that success in introducing change needs education and training at all levels of the company. People have to be managed and encouraged to develop new ways of working. Processes and systems have to be simplified. There has to be involvement at all levels from the top of the company to the shop floor. Within such a framework objectives can be agreed with groups and individuals. This whole activity becomes 'people empowerment' which may be uncomfortable to managers brought up in the Taylor tradition of managers planning and workers doing.

A pre-condition of integration is an understanding of how an activity is carried out and how it relates to the other activities that are to be integrated. This applies whether it is process integration, systems integration or methods integration. A lot of attempts at integration have failed through trying to integrate a poorly understood and confused activity by the use of computer hardware and software. The result has been automated chaos. The key to success is to simplify and understand the process, and then to integrate.

The importance of this understanding can be seen in examining the comparatively simple process which Hepworth established as a computer integrated manufacturing operation. The systems and methods for integrating this well-understood manufacturing process resulted in the complex of relationships shown in Hepworth, Case Study 3, Figure CS3.1. Each system had to have parameters and rules established. Common or interrelated data bases had to be established so that computers could be programmed to control the process.

3.5.4 Integrating horizontally

This section considers horizontal integration in the company but this integration, inevitably, spills over into the supply chain. Integration takes place when the manufacturing activity as defined in Figure 3.5 has a common goal with marketing. A foundation for this is Simultaneous Engineering (SE).

Simultaneous engineering

Simultaneous engineering is the activity that takes the needs of the customer and transforms them into a product. The product has to satisfy customer needs in terms of function and cost. The product must also satisfy the company in terms of profit contribution.

To achieve these aims product design must make use of the expertise of the supply chain to provide materials and components. Inside the company the supply chain becomes the stages by which the raw materials and components are transformed into the customer's product. To produce a design that is satisfactory to the customer and to the supply chain a team representing those interests must be assembled. The team needs the authority to arrive at decisions without constant reference to higher levels for ratification. The team may need to include suppliers, i.e. personnel, who are not company employees. The team will be multi-functional and will not need all members throughout the evolution of the design. Members will move in and out of the team as the design requirements vary. Rolls-Royce, Case Study 5, Figure CS5.13, presents the variation in a SE team at Rolls-Royce.

The objective is to satisfy the needs of the design definition. The design

Strategy, Integration and Focus

activity takes place in a commercial environment that can affect the agreed introduction as shown in Case Study Extract 3.8.

Stanley Tools integrated the process of manufacture with the design of a product to satisfy the needs of the customer. Commercial considerations altered the time scale of the product introduction. Emergency action of this magnitude is a response that companies have to make to changes in the environment.

Product design is a starting point for horizontal integration. The design

Case Study Extract 3.8 Stanley Tools – modified introduction of the Magnum screwdriver (Case 6)

Stanley fixed blade screwdriver sales had been declining in a fairly stable market. The decline was due to lack of a product with features which would enable it to occupy the top place in the screwdriver market. With such a product Stanley could re-price its existing range.

The Magnum screwdriver introduction was initiated by a multi-function management team established to develop screwdriver business. This is a normal way of development at Stanley using product line management teams. These teams control the product line and are a variant of the SBU and profit centre concept.

The original phasing of the project recognised that the protracted leadtime required for manufacture of injection moulding tools, pre-delivery testing and de-bugging were the main factors dictating the launch date. In view of this and the subsequent launch build-up, the earliest stock date was to be mid-May 1991.

The following key dates applied with regard to the moulding machine and tooling in order to achieve a UK launch date of June 1991:

Submit business plan	September 1989
Place capital orders	November 1989
Initial tool trials (Germany)	September 1990
Delivery of moulding machine and tooling	November 1990
Installation	December 1990
In production	January 1991
In stock	End May 1991
Follow up launch plans to be:	
France	September 1991
Northern region and Italy	January 1992

The deteriorating competitive position of the Stanley range of screwdrivers required that as soon as the project became firm in November 1989 one year had to be eliminated from the timetable. This was achieved and launch took place in late 1990. The result has been an increase in unit sales of 12 per cent with the value of turnover increasing by 44 per cent.

fixes the quality standards needed to achieve customer satisfaction. SE ensures that those standards can be met throughout the supply chain. Management of the supply chain will ensure that design parameters can be achieved in practice.

Quality

Quality is an integrating factor in the organisation. Quality must be defined so that successive stages in the supply chain can contribute to the final product with the confidence that the product quality will satisfy the customer. Design for manufacture ensures that process capability will give right-first-time manufacture. Unit batching becomes a reality. Under these conditions work flow can be organised.

Work flow organisation

The horizontal co-ordinating mechanism on the shop floor is a system of work flow that will make product to customer demand. Customer demand, forecast or actual, is broken down into a series of dependent demands. These dependent demands are the levels in the Bill of Materials (BOM) that have either to be manufactured or procured. The overall leadtime of this schedule determines at what time the stages of procurement or manufacture must commence in order to satisfy the promised delivery to the customer.

A production process can be designed to satisfy customer demand. As described in Rolls-Royce, Case Study 5, group technology can be used to group together parts similar in shape, size or method of manufacture. These grouped products can be manufactured in a cell. Fisher Controls, Case Study 2, re-arranged their work flow from a process flow to a product flow. In both cases the grouping of similar machines that forms the basis of process flow was altered to a grouping which gives product flow. Kodak, Case Study 4, Figure CS4.1 shows product flow, dependent demand (BOM) and kanbans.

The grouping of machines to satisfy product flow rather than process flow means that management must recognise that satisfaction of customer orders is more important than machine or departmental efficiencies. These groupings of machines and processes form manufacturing cells that are manned by teams.

Cells and teams have been located between horizontal and vertical integration in Figure 3.12 because they serve both purposes. They integrate horizontally as described. In respect of the strategic aim of the company and its integration vertically, cells and teams are the operational level at which strategy is implemented.

Manufacturing systems engineering

A particular application of the cell concept was developed by Lucas Industries as manufacturing systems engineering (MSE). MSE is designed to integrate

manufacturing systems. Parnaby (1988) gives a five-step approach:
1. Data collection on markets, products and manufacturing processes.
2. Engineering analysis to define volume–variety groups.
3. Definition of a business architecture forming SBUs and cells.
4. Testing and refining the architecture by what-if questioning.
5. Definition of the information flows needed.

The approach concentrates on gaining major improvements from the facilities which exist. Only after the benefits have been gained from these improvements should capital expenditure on new technology be considered. The emphasis then should be on improvement of the total system, not on piecemeal improvement.

■ 3.6 Conclusions

The success of a manufacturing company depends upon its ability to integrate activities in order to focus on and exploit market opportunities. The strategy for exploiting market opportunities is driven by corporate, marketing and manufacturing strategies.

Manufacturing strategy needs to formulate and implement coherent actions in all areas from receipt of customer demand to satisfaction of that demand. Much manufacturing activity is located outside the company either as bought in material or as the sub-contract manufacture of parts and assemblies. Manufacturing strategy must embrace these areas and set policies. Overall manufacturing strategy must dovetail with marketing strategy to form the basis for competitive advantage.

For companies making diverse products the adoption of an organisation structure based on SBUs and profit centres aids strategy formulation, implementation and control. Operational autonomy results and this increases accountability for performance. SBUs can be highly motivating and give rapid response to changes in the environment.

To gain the synergistic advantages of being an integrated company rather than a federation it is necessary to overlay the SBU/profit centre structure with a co-ordinating architecture.

In the manufacturing company decisions can be made operationally which pre-empt the strategy of the company. To prevent this occurring functional boundaries must be destroyed even if functions continue to exist. Managers have to be educated to understand that plant replacement, skills development and sourcing decisions are strategic as well as operational. To be aware of the implications of such decisions managers need to be trained and developed beyond their functional experience. Responsible experience in other areas of the business is required.

Integration of manufacturing is important to enable a manufacturing company to make a rapid response to customers' needs. Rapid response is needed to meet changes in the volume and mix of products. For all companies rapid response by manufacturing to changes in the business environment is important. For those companies where product life cycles are shortening, rapid response in the introduction of new products is vital.

Integration applies to process, systems and methods both manual and mechanised. Integration has to take place in the company but it also has to take place with customers and suppliers. Integration of the supply chain is needed if the company is to gain competitive advantage.

Integration depends on simplifying the activities of the company to gain an understanding of the complex relationships which obtain. As these activities are simplified relationships can be clarified. The mechanism of integration itself may be simple. For example, JIT is a very effective way of gaining integration while at the same time being simple. It is simple because it does not need computers and complex control systems. It is not, however, simple to design and execute.

Integration involves all company employees but it involves them in groups that can identify how they can make a contribution to the goals of the company.

Integration depends on the ability to implement the changes that have been identified. Designs of structures and systems offer the opportunity for integration. Integration does not take place until these structures and systems are made to work by people. The experience with MRP is that more than half the systems do not do what was expected. The fault may be with the systems but people must have a wider view of implementation than system introduction.

4 Quality

■ 4.1 Introduction

Quality has been the competitive factor that has enabled Japanese manufactured goods to establish a dominant position in large areas of Western durables markets from the 1970s. The Western response has been to attempt to reach similar quality standards. The development of quality is put into an historical perspective in Figure 4.1, and a number of approaches to quality improvement are summarised in Table 4.1.

As can be seen from Figure 4.1 the development of quality in Japan, in terms of formal national quality initiatives, has a lead of some 20 years compared to America which in turn is ahead of the EU. Miller, De Meyer and Nakane (1992) identified that quality became a driving theme so suddenly in American manufacturing industry in the 1980s that they termed it a 'quality revolution'. The analysis from their manufacturing futures survey was that, by the end of the 1990s, there was a relentless drive towards better quality continuing throughout the world. This drive continued in Japan and the United States, and was gaining strength in Europe.

During the 50 years shown in Figure 4.1 the Japanese have developed product quality to a very high level. This has been made possible by a very careful definition of customer needs, rigorous control of the manufacturing process and improvement in quality of incoming supplies. Thus, customer satisfaction has been obtained by effective management of the whole of the supply chain.

The UK, in common with the rest of Western manufacturing, is in a catch up situation. It is not only product quality that needs to be improved but the wider aspects of quality variously termed Total Quality Management (TQM), Company-wide Quality Control (CWQC) or World Class Manufacturing.

The approaches to quality improvement shown in Table 4.1 are quite prescriptive. Companies must make a choice of an approach suited to their needs and then be committed to that choice. The four approaches shown are

The costs of quality

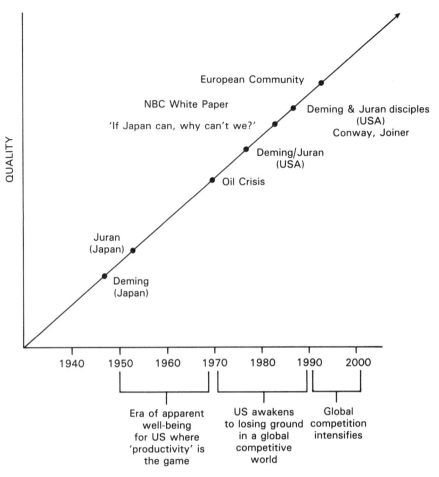

Figure 4.1 Global quality awareness (*Source*: Quality Leadership Process, Kodak Ltd)

not the only ones available. They have been presented to show readers some of the differences in ways of achieving TQM. Each of these approaches has resulted in success.

TQM improves quality. In addition, by the elimination of waste in the supply chain, the cost of manufacture can be lowered. Thus, TQM should result in reduced product cost.

■ 4.2 The costs of quality

The costs of quality depend upon how they are defined. Dale and Plunkett (1991) are critical of BS 6143 *Guide to the Determination and Use of Quality*

Table 4.1 Approaches to quality improvement

Deming's 14 points for management	Juran's 10 steps to quality improvement	Conway's 6 tools for quality improvement	Crosby's 14 steps to quality improvement
1. Create constancy of purpose for improvement of product and service 2. Adopt the new philosophy 3. Cease dependence on inspection to achieve quality 4. End the practice of awarding business on the basis of price tag alone. Instead, minimise total cost by working with a single supplier 5. Improve constantly and forever every process for planning, production and service 6. Institute training on the job 7. Adopt and institute leadership 8. Drive out fear 9. Breakdown barriers between staff	1. Build awareness of the need and opportunity for improvement 2. Set goals for improvement 3. Organise to reach the goals (establish a quality council, identify problems, select projects, appoint teams, designate facilitators) 4. Provide training 5. Carry out projects to solve problems 6. Report progress 7. Give recognition 8. Communicate results 9. Keep score 10. Maintain momentum by making annual improvement part of the regular systems and processes of the company	1. Human relations skills – responsibility of management to create at every level, among all employees, the motivation and training to make the necessary improvements in the organisation 2. Statistical surveys – the gathering of data about customers (internal as well as external), employees, technology and equipment, to be used as a measure for future progress and to identify what needs to be done 3. Simple statistical techniques – clear charts and diagrams that help identify problems, track work flow, gauge progress, and indicate solutions	1. Make it clear that management is committed to quality 2. Form quality improvement teams with representatives from each department 3. Determine where current and potential quality problems lie 4. Evaluate the cost of quality and explain its use as a management tool 5. Raise the quality awareness and personal concern of all employees 6. Take actions to correct problems identified through previous steps 7. Establish a committee for the zero defects program 8. Train supervisors to actively carry out their part of the quality improvement program

4. Statistical process control – the statistical charting of a process, whether manufacturing or non-manufacturing, to help identify and reduce variation
5. Imagineering – a key concept in problem solving, involves the visualisation of a process, procedure, or operation with all waste eliminated
6. Industrial engineering – common techniques of pacing, work simplification, methods analysis, plant layout and material handling to achieve improvements

9. Hold a 'zero defects day' to let all employees realise that there has been a change
10. Encourage individuals to establish improvement goals for themselves and their groups
11. Encourage employees to communicate to management the obstacles they face in attaining their improvement goals
12. Recognise and appreciate those who participate
13. Establish quality councils to communicate on a regular basis
14. Do it all over again to emphasise that the quality improvement program never ends

10. Eliminate slogans, exhortations and targets for the workforce
11. Eliminate numerical quotas for the workforce and numerical goals for management
12. Remove barriers that rob people of pride of workmanship. Eliminate the annual rating or merit system
13. Institute a vigorous program of education and self-improvement for everyone
14. Put everybody in the company to work to accomplish the transformation

Source: Quality Leadership Process, Kodak Ltd

Quality

Related Costs which they regard as an abridged and poor imitation of the ASQC *Quality Costs – What and How*. In general they see the costs of quality as the costs of:

- Prevention.
- Appraisal.
- Failure.

These costs are high as two examples show:

1. In 1978 the UK government estimated them to be £10,000 million or 10 per cent of GNP.
2. In 1985 NEDO estimated that 10–20 per cent of an organisation's total sales value is accounted for by quality related costs. Using the figure of 10 per cent it is estimated that UK manufacturing could save £6 billion annually.

Ninety-five per cent of quality costs are expended on appraisal and failure and are open to saving by implementing TQM. However, the lead given by the UK government in setting quality standards has been the procedural implementation of quality (BS 5750) rather than TQM. This procedural approach has been adopted by the EU (ISO 9000).

■ 4.3 Quality standards

Governments have public procurement programs where they want to be assured that manufacturers are adhering to procedures aimed at guaranteeing standards of quality. This has led to quality certification under BS 5750/ISO 9000 standards. To be considered as potential suppliers for government contracts certification under these standards is a necessity. With the gradual elimination of barriers in the EU the possession of such certification will increase in importance as customers expect suppliers to have such certification. However, as many companies have found, such certification does not alter the technology in the supply chain and does not become TQM.

On the other hand, Ford approached standards within their supplier network with a quality system, Q101 which aimed at improving the technology of manufacture. Thomas (1987) describes the system under three main headings:

1. Advanced quality planning.
2. Quality execution.
3. Performance monitoring.

The tool that underpinned the system was statistical process control (SPC). 'What is new is that not only can these techniques be used to achieve defect prevention rather than detection, but that they can be used to achieve an ongoing improvement to quality and productivity'.

Garvin (1988) identifies a progression in the control of quality from inspection through statistical quality control, quality assurance to strategic quality management. This progression is from a situation where customer quality may be met but at a considerable cost to the company to a situation where customer quality will be met at a reduced cost to the company.

The goal is TQM where quality is 'managed in'. The 'managing in' of quality has now been recognised by the British Standards Institution with their BS 7850:1992, Total Quality Management.

Figure 4.2 represents the thrust of the standard:

- Systems. These are considered to be initial building blocks in moving towards a TQM environment. The various Charter initiatives fit into this segment.
- People. A total quality organisation has to capture the hearts and minds of all company personnel. The Investors in People Initiative is expected to be a driving force with the Management Charter Initiative and National Vocational Qualifications spanning 'systems' and 'people'.
- Quality improvement methods. This is the theme of Part 2 of the standard.
- The rest – 'a catch-all sector' – recognises that different companies have different needs.

This standard is written to include all organisations not just manufacturing.

The problem for British manufacturing companies is that the need for improvement in quality is urgent. Where should a company start?

Figure 4.2 Total quality environment (*Source*: Sanderson, *BSI News*, November 1992)

Quality

Japan, by the 1990s, has some 40 years' experience in developing quality. America has some 20 years and Britain perhaps 10, although the development is highly variable across British manufacturing. The Japanese transplants get off to a flying start. Quality standards and practices are already in existence in the parent company. The operation is often established on a greenfield site where quality standards and practices can be set up integral to starting production.

While the need to improve quality is urgent in British brownfield manufacturing operations, the companies have to do so while under pressure to retain or increase market share, to make profit, to survive.

This chapter will discuss TQM to help management make choices which will lead to continuous improvement in quality in their companies. It must be emphasised that working towards TQM is not a question of seizing on a set of rules, e.g. from those shown in Table 4.1, but of identifying how TQM is to be achieved in the particular company.

A starting point is the customer.

■ 4.4 The customer

The customer has needs that must be satisfied by the company's product. Not only has the product to satisfy the customer's needs, it also has to satisfy them, in some way, better than competing products. It may be argued that product quality is only one aspect of customer satisfaction and that functionality, features, price, etc. are equally or even more important. However, TQM by controlling quality throughout the company's activities determines the 'competitive package' that will give the company a competitive edge in satisfying customer needs.

The design of a product that incorporates customer perspectives as identified by the customer and the company has limitations. A narrow perspective may lead to insufficient consideration being given to competitor solutions. Competitive benchmarking is a technique for solving this problem and will be examined in Section 4.6.

To satisfy customer needs it is important that those needs are known throughout the company and the supply chain. Design has to involve a wide group of people, for example, marketing, production, product development, etc. One technique that has been developed to identify customer needs and guide the product development process is Quality Function Deployment (QFD).

QFD uses a planning matrix to identify the customer needs on the basis of 'what' those needs are and 'how' they can be satisfied. These matrices take the desired product attributes, convert them into design parameters and ultimately into production requirements. Thus QFD carries the 'voice of the customer' across the organisation's functions such as design, process engineering, manufacturing and into the supply chain.

Nichols (1990) states that QFD has readjusted the approach by design teams at Digital Equipment Co. Ltd. As a result of the use of QFD the team:

- Goes out and listens to customers.
- Gains a clearer understanding – debate is reduced.
- Has greater competitive awareness.
- Concentrates on design target values rather than acceptable limits.

It is important to consider the supply chain in the context of product design and of TQM. As has already been stated, on average 50 per cent of manufactured product is bought in. This raises two issues:

1. For components wholly designed in company but sourced outside there has to be an effective mechanism to ensure that the supplier is capable of meeting that specification. Also there must be a procedure to ensure that components received from the supplier conform to the specification. This aspect will be examined further in Section 4.10.
2. Components or sub-assemblies may be designed by suppliers because of their expertise, e.g. in the automotive industry Automotive Products design braking systems and Britax Vega design lamp clusters for specific models of cars. This design by the supplier may arise because they possess a specific technology that the manufacturer of the product either has never had or has made a strategic decision to relinquish.

The supplier–customer relationship that obtains between a company and its external suppliers has been developed within manufacturing companies. Each successive operation becomes the customer of the previous operation giving a series of supplier–customer relations from receipt of raw material to completion of the product. Thus it can be seen that this supplier–customer relationship extends throughout a company's supply chain.

In order for these multiple supplier–customer relationships to work effectively it is necessary, at each stage, to be able to identify quality specifications and to put into place methods of ensuring that value being added by subsequent operations is to a 'perfect' product. Zero defects and right-first-time must be achieved if perfect products are to flow from suppliers to customers.

4.4.1 The importance of QFD

Miller, De Meyer and Nakane (1992) give details of company responses to a questionnaire in 1990 asking which of 26 programs and activities have given pay-offs in the last two years, and which will receive emphasis over the next two years. The data given has been analysed and Tables 4.2 and 4.3 have been derived therefrom.

Quality

Table 4.2 Past pay-offs from programs and activities by country

	US	UK	Japan
1	Manufacturing reorganisation	Linking manufacturing strategy to business strategy	Management training
2	Interfunctional work teams	Worker training	Developing new processes for old products
3	Statistical quality control	Supervisor training	Developing new processes for new products
4	Linking manufacturing strategy to business strategy	Developing new processes for new products	Quality circles
5	Just-in-time	Statistical quality control	Computer aided design
6	Computer aided design	Manufacturing reorganisation	Quality function deployment
7	Reconditioning physical plants	Management training	Value analysis/product redesign
8	Supervisor training	Quality function deployment	Investing in improved production–inventory control systems
9	Developing new processes for new products	Investing in improved production–inventory control systems	Integrating information systems in manufacturing
10	Worker training	Interfunctional work teams	Supervisor training
11	Value analysis/product redesign	Reconditioning physical plants	Linking manufacturing strategy to business strategy
12	Developing new processes for old products	Integrating information systems across functions	Worker training
13	Closing and/or relocating plants	Integrating information systems in manufacturing	Statistical quality control
14	Quality function deployment	Giving workers a broad range of tasks and/or more responsibility	Giving workers a broad range of tasks and/or more responsibility
15	Management training	Developing new processes for old products	Interfunctional work teams
16	Computer aided manufacturing	Computer aided design	Integrating information systems across functions
17	Design for manufacture	Hiring in new skills from outside	Robots
18	Giving workers a broad range of tasks and/or more responsibility	Value analysis/product redesign	Manufacturing reorganisation
19	Quality circles	Computer aided manufacturing	Design for manufacture
20	Hiring in new skills from outside	Just-in-time	Just-in-time
21	Investing in improved production–inventory control systems	Quality circles	Reconditioning physical plants
22	Integrating information systems across functions	Flexible manufacturing systems	Computer aided manufacturing
23	Flexible manufacturing systems	Design for manufacture	Flexible manufacturing systems
24	Integrating information systems in manufacturing	Closing and/or relocating plants	Hiring in new skills from outside
25	Activity based costing	Activity based costing	Activity based costing
26	Robots	Robots	Closing and/or relocating plants

Source: Derived from *Benchmarking Global Manufacturing*, J.G. Miller, A. De Meyer and J. Nakane, Business One Irwin, 1992

The customer

Table 4.3 Future emphasis on programs and activities by country

	US	UK	Japan
1	Linking manufacturing strategy to business strategy	Linking manufacturing strategy to business strategy	Integrating information systems in manufacturing
2	Giving workers a broad range of tasks and/or more responsibility	Worker training	Developing new processes for new products
3	Worker training	Supervisor training	Investing in improved production–inventory control systems
4	Statistical quality control	Interfunctional work teams	Linking manufacturing strategy to business strategy
5	Supervisor training	Integrating information systems across functions	Developing new processes for old products
6	Interfunctional work teams	Quality function deployment	Integrating information systems across functions
7	Management training	Statistical quality control	Computer aided design
8	Quality function deployment	Management training	Quality function deployment
9	Integrating information systems in manufacturing	Integrating information systems in manufacturing	Management training
10	Integrating information systems across functions	Developing new processes for new products	Design for manufacture
11	Developing new processes for new products	Giving workers a broad range of tasks and/or more responsibility	Value analysis
12	Design for manufacture	Just-in-time	Robots
13	Just-in-time	Manufacturing reorganisation	Supervisor training
14	Computer aided design	Investing in improved production–inventory control systems	Computer aided manufacturing
15	Investing in improved production–inventory control systems	Developing new processes for old products	Quality circles
16	Value analysis/product redesign	Computer aided manufacturing	Flexible manufacturing systems
17	Manufacturing reorganisation	Flexible manufacturing systems	Giving workers a broad range of tasks and/or more responsibility
18	Computer aided manufacturing	Reconditioning physical plants	Worker training
19	Developing new processes for old products	Activity based costing	Just-in-time
20	Flexible manufacturing systems	Quality circles	Reconditioning physical plants
21	Activity based costing	Hiring in new skills from outside	Interfunctional work teams
22	Hiring in new skills from outside	Computer aided design	Manufacturing reorganisation
23	Reconditioning physical plants	Design for manufacture	Statistical quality control
24	Quality circles	Value analysis/product design	Hiring in new skills from outside
25	Closing and/or relocating plants	Closing and/or relocating plants	Activity based costing
26	Robots	Robots	Closing and/or relocating plants

Source: Derived from *Benchmarking Global Manufacturing*, J.G. Miller, A. De Meyer and J. Nakane, Business One Irwin, 1992

Quality

Table 4.4 Importance of QFD

	Ranking of 26 programs and activities		
	US	UK	Japan
Pay-off in the past	14	8	6
Future emphasis	8	6	8

1 = most important 26 = least important
Source: Derived from *Benchmarking Global Manufacturing*, J.G. Miller, A. De Meyer and J. Nakane, Business One Irwin, 1992

The importance attached to QFD by companies in the US, UK and Japan is shown in Table 4.4. As can be seen, QFD is considered to be important by companies in all three countries.

However, QFD was developed in Japan in the 1970s. Analysis of Table 4.4 suggests that QFD, as with many other quality related activities, has already given the Japanese a huge pay-off. While the future emphasis of QFD is moving up the ranking in the US and UK, it is moving down in Japan.

The US and UK, in their attempt to improve quality, increase the future emphasis on many quality related functions. The emphasis in Japan is moving from quality towards activities that will give them flexibility, their future competitive weapon.

■ 4.5 Quality planning

Quality planning is a strategic activity. Garvin (1988), in considering strategic quality management, puts emphasis on the market and consumer needs. Focusing on these needs is the starting point for quality planning. In order to plan for quality there has to be an effective way of capturing the 'voice of the customer' and making it heard throughout the company and its supply chain. The goal of the business must be the satisfaction of its customers. One way that this goal can be expressed is to place the assets of the business at the disposal of the customer to the mutual benefit of the customer and the company.

Quality planning to realise customer satisfaction will be specific to the company and its customers. What TQM is and how it is to be achieved has to be planned and developed by each company. Quality planning aligns all quality objectives throughout the organisation. However, it is possible for management to think that it has effective quality planning only to find that there is failure in communication and implementation.

De Vries and Rodgers (1991) made an analysis of what went wrong with the Philips 1983 company-wide quality improvement (CWQI) campaign and their findings are summarised in Illustration 4.1. The problems identified in Philip's quality planning are frequently encountered in manufacturing companies. The problems of organisational boundaries and functional rivalry have been discussed in Chapter 3. Integration and focus are needed for a company

Illustration 4.1 Problems with Philips 1983 CWQI campaign

- Quality initiatives did not cross organisational boundaries.
- Employees were mobilised in motivational programmes BUT common targets were not set.
- The customer remained invisible. Teams were trained in quality tools. The programme aimed to improve quality without defining quality goals.
- Functional rivalry. The initiative was involved in functional rivalry – other developments in logistics, industrial engineering and human resources competed.
- The importance of normal work achievement was separated from quality improvement achievement. Functional performance indicators continued to be used to judge operational performance.

Quality performance was judged in quality team meetings.

Source: J. De Vries and L. Rodgers, Bridging business boundaries, *The TQM Magazine*, December 1991

to be effective in competing. SBUs are being used as a potent way of integrating manufacturing and focusing the business on customer needs. The invisibility of customers and the difficulty of judging quality improvements can be addressed by:

- Simultaneous engineering – see Chapter 3.
- People empowerment – see Chapter 5.
- Target costing – see Chapter 6.

However, in deciding to improve customer visibility, it is important to take account of the differences imposed by products and markets. For example, DSF, Case Study 1, has an approach to quality where the customer is important but individual products are not the focus, see Section CS1.6 and Figure CS1.5. Stanley, Case Study 6, focuses round products as demonstrated in Section CS6.1 describing the launch of the Magnum screwdriver.

The design, production and launch of a new car take place at discrete intervals of time, say five years. Fisher Controls, Case Study 2, on the other hand, provides unique process control solutions for their customers. This results in the need for constant enhancement of their PCB functionality. Design and production have to react to a constantly changing customer need.

Quality planning is the basis for realising TQM. In practice quality planning will take place differently in different organisations as shown in Illustrations 4.2 and 4.3. A significant feature of planning in both illustrations was the top management commitment. The issues raised by actively moving to TQM are important. The Philips example shows the problems that can arise if they are not addressed.

The policies adopted in the illustrations were different. Nissan was aiming at a standard already achieved in Japan. The planning was about how to reach

ty

> **Illustration 4.2** Quality – above all: TQM at Nissan
>
> The transplant aspect is very strong. The fact that a standard had been seen in Japan brought about a determination to equal that standard and then beat it. An uncompromising approach was made where quality would not be subordinated to production schedules. The transfer was to a British culture where it was recognised that slogans and posters would not contribute to TQM.
>
> The NMUK agreement with the AEU states: 'both parties are agreed on the need to establish an enterprise committed to the highest levels of quality, productivity and competitiveness using modern technology and working practices and to make such changes to this technology and working practice as will maintain this position'.
>
> The Constitution of the Company Council states that it will be involved in 'matters concerning the Company's business, e.g. quality, production levels, market share, profitability, investment etc.'.
>
> The Company's philosophy statement issued to all employees: 'We aim to build profitably the highest quality car sold in Europe'.
>
> Putting TQM into practice meant:
>
> - Top management commitment.
> - Supervisor leadership on resolving quality problems.
> - Training employees to a standard where they can achieve quality.

Source: P. Wickens, *The Road to Nissan*, Macmillan, 1987

those standards in the UK. An integral part of NMUK strategy was that UK workers had experience of working in the Japanese parent factory. Kodak had a company-wide vision of quality improvement. The planning policy to fulfil this vision is based on the Deming cycle, underpinned with extensive training and techniques. The driving force for change in Kodak in the UK had been their introduction of MRPII/JIT.

■ 4.6 Competitive benchmarking

Benchmarking is the continuous process of measuring products and practices against the toughest competitors, or those companies regarded as market leaders in related processes.

Competitive benchmarking as a part of QFD seeks a wider solution to the satisfaction of customer needs by making comparison with competitor solutions. QFD sets out to satisfy customer needs. In the process of benchmarking the company must identify how the competition satisfies those needs. This is not new. The automotive industry for years has torn apart competitors' models to identify cost and means of manufacture: 'reverse engineering'.

Illustration 4.3 Kodak quality leadership process

A call to action

Continuous improvement of our products, services and operations is our vital link to a successful future. I call upon each of you to make a difference through your personal part in the quality leadership process.

Kay R. Whitmore
Chairman, President and Chief Executive Officer

Continuous improvement cycle

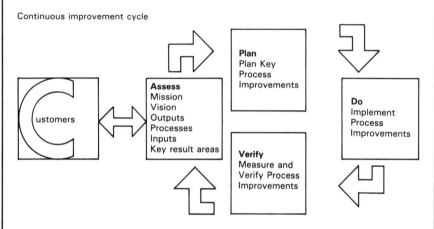

Customers
  Identify customers
  Discover customer needs

Assess
  Develop mission
  Develop vision
  Identify outputs supplied to customers
  Identify process to supply outputs
  Determine inputs and suppliers
  Identify key result areas and measures

Plan
  Identify improvement opportunities
  Select major improvement opportunities (MIOs) linked and aligned to your KRAs
  Develop a plan for managing the improvement opportunities (MIOs)
  Initiate the improvement plan and enable resources

Do
  Diagnose the existing process
  Eliminate causes of problems identified for existing processes
  Diagram new/improved process
  Develop measure of the new/improved process
  Communicate with those affected by the change
  Document the improved process
  Standardise the improved process

Verify
  Review charts data
  Identify out-of-control points
  Plan investigative/corrective/capturing actions for out-of-control performance
  Reinforce team member contributions and celebrate team progress
  Plan additional reinforcement to be given
  Review and modify MIO action plan and reinforcement plan
  Review project team progress
  Review and plan additional action items

Source: Kodak internal company publication

Quality

If benchmarking is only used for being as good as the competitor, leap-frogging will take place. The ultimate winner will be the company that uses the benchmark data to come up with an innovative design for the customer demand.

Competitive benchmarking aims to measure performance throughout the company.

- The ultimate measure is made by the customer. The customer's perception of the performance of the product must be the goal to satisfy.
- To reach this goal comparison must be of process as well as output.
- It is not sufficient to compare process performance to that of immediate competitors. Comparison has to be with the best in the class, for example the best distribution system may not be that of a competitor but of a company manufacturing different products, or even a company in a different industry.

Miller, De Meyer and Nakane (1992) have used their manufacturing futures survey as a means of benchmarking countries. Figure 4.3 shows the performance improvements made between 1988 and 1989 for eight factors.

Companies can use these benchmarks of average rates of improvement to identify their own level of performance. It must be remembered that these are average rates of improvement and that there are significant differences in the starting point, for example, both the US and Europe show better percentage improvements in quality than Japan but this is only a catch up situation. Japan still has a tremendous lead in quality as in many of the other activities considered.

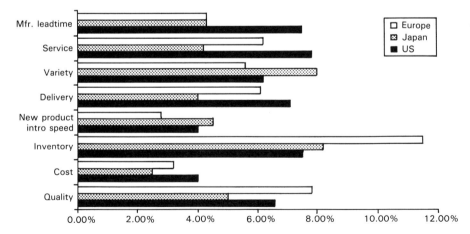

Figure 4.3 Performance improvements, 1988–1989 (*Source*: Miller, De Meyer and Nakane (1992))

Competitive benchmarking

This lead can be seen in the pay-off and future emphasis on statistical quality control (SQC), see Table 4.5. The US and UK are getting high pay-offs from SQC. Japan, which had started using the activity in the 1950s was seeing a lower pay-off. As a result the future emphasis on SQC in Japan was very low whereas for the US and the UK it was high and would continue to be an important activity in their continuing struggle to reach the quality standards already accepted in Japanese manufacturing companies.

4.6.1 The practice of benchmarking

The steps involved in benchmarking are shown in Figure 4.4. Benchmarking is a process which as it is developed and refined will give the company potential for improving its competitive stance.

Table 4.5 The importance of SQC

	Ranking of 26 programs and activities		
	US	UK	Japan
Pay-off in the past	3	5	13
Future emphasis	4	7	23

1 = most important 26 = least important
Source: Derived from *Benchmarking Global Manufacturing*, J.G. Miller, A. De Meyer and J. Nakane, Business One Irwin, 1992

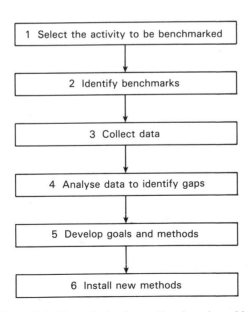

Figure 4.4 Steps in implementing benchmarking

Quality

Fisher Controls, Case Study 2, made a comparison with competitors although they did not have a system which they called benchmarking. As can be seen from Case Study Extract 4.1, they adopted the approach that is needed to make a success of benchmarking.

The crux of the Fisher approach, as of all benchmarking, is not to identify what improvements can be made given current company practices but what improvements are needed to reach or surpass the benchmark levels of performance. Companies have to examine ways in which existing processes, systems and management philosophies can be improved to give the needed results. If the results cannot be obtained then new processes, systems and management philosophies must be adopted to reach the required level of performance. Benchmarking in its examination of competitor activities gives the opportunity for this examination and analysis.

Benchmarking sets the levels for effective competition. If a company cannot or will not reach these levels future results will be unsatisfactory and survival will be jeopardised. Having identified the goal, plans have to be made and implemented for the company to reach that level of performance.

The preliminary questions to be asked at the start of benchmarking are:

- Is there a competitor who is better?
- What is the gap between the company and the competitor?
- What are they doing which gives them better performance?
- How can this company learn to perform as well?
- How can we put this learning into practice?

Case Study Extract 4.1 Fisher Controls (Case 2)

1	People involvement
2	Total quality
3	Supplier partnering
4	MRP/JIT
5	Loss prevention

The question under each of these headings was 'where should we be in two to five years?' Targets were set based on what was needed in comparison with competitors' achievements and on what the company identified as necessary, e.g. a 60 per cent reduction in inventory holding.

At this stage management did not know if the targets could be achieved but they accepted that these were the targets which had to be met if Fisher was to be competitive.

Returning to an examination of the steps of benchmarking, there is a need to select and focus on specific activities, identify benchmarks, collect and analyse data and develop goals and methods and implement them.

Select activities

Formulating strategy as outlined in Chapter 3 will focus on areas where the company is seeking to gain sustainable competitive advantage. Benchmarking is a fundamental part of QFD and, in responding to the 'voice of the customer' activities for benchmarking will become evident.

The search for competitive advantage will reveal areas within the process, the supply chain and administration where improvement is needed. Porter's value chain (1985b) can be used to identify key areas where value can be added or steps can be taken to eliminate activities that are failing to add value.

Pareto analysis, the 80/20 rule, should be used to make sure that the activities chosen give the best pay-off.

Identify benchmarks

Identification of the class leader will be closely related to the selection of activities. Effective strategy formulation will have identified competitors. Benchmarking is an attempt to sharpen the company reaction by making the total company, not just a planning group, aware of the need to improve performance. Benchmarking needs to pervade the company so that it becomes an activity involving everyone.

Where benchmarking is not concerned with the product but with other activities the selection can be guided by company personnel with expertise in the area and, therefore, with knowledge of 'best in the class' companies. These companies need not be competitors, and frequently will not be. If the companies are not competitors access becomes easier.

Collect data

Company employees are a valuable source of information. Traditionally sales forces have been regarded as sources of information on competitor products. However, accountants, production controllers, purchasers, quality controllers, etc. belong to professional bodies which publish articles, hold meetings and promote research. Here is a valuable source of information on practices and processes.

While sales forces may be a valuable source of customer information, their information is liable to be biased. There are other sources of corroboration or contradiction of the sales force's view such as service engineers and customer complaints departments. Meetings with customers by senior company personnel are used by companies as sources of information.

Quality

Where products are distributed to the final customer through chains, e.g. retail distribution, it is important that companies set out to obtain information from suppliers and distributors and the ultimate customer. Illustration 4.4 shows how a small company has set out to collect information both on its own and its competitors' products.

Published information such as government statistics and company reports are also valuable sources of information.

Analyse data to identify gaps

Analysis has to be of methods and processes not results. The objective is to establish the gap between company performance and benchmark performance. The important activity is to analyse reasons for the gap in performance and how that gap can be closed. It is possible that the benchmark company will be willing to exchange information or even permit a visit by a company team. As suggested in Illustration 4.4 suppliers may be sources of information. Trade exhibitions, professional bodies and published material are sources of information as to how methods and processes can be improved to reach the benchmark standard.

Illustration 4.4 Interior Stone Features Ltd

Interior Stone Features Ltd (ISF) is a small company with a turnover of £2m per annum which, as a part of its activity, supplies mantels and marble sets to five of the twelve regions of British Gas. British Gas subsequently install these mantel sets as surrounds to the flame effect fires that they sell.

- Benchmark data is available to ISF sales staff by identifying competitor prices and the standard of their products which are also on display in Gas Board showrooms.
- How the mantel sets are constructed and what materials have been used can be found out by purchasing competitors' products and stripping them down.
- Showroom staff are a source of information on customer choice, their likes and dislikes.
- ISF staff talk to Gas Board fitters to find out relative advantages and disadvantages of installation.
- ISF's suppliers of wood and marble are able to comment upon manufacturing practices adopted by competitors, as are tool suppliers. Particularly important are the paint and varnish suppliers. The finish of the mantels is an important factor in customer choice. In the first place the visual appearance in the showroom attracts the customer. Subsequently, since the mantels are subject to temperature variation, resistance to cracking of the paint/varnish is important to continued customer satisfaction.

Brainstorming on how to close the gap will unlock ideas existing within the company. Value engineering/analysis is a technique that can be used to find reasons for the gap in performance.

Develop goals and methods

Goals, methods and implementation are closely related. If the goals are to be the standards set by the benchmark, methods of process improvement must be determined. There still remains a problem of obtaining the commitment to the process improvement and the achievement of the goals.

Involvement of all company personnel is vital to benchmarking. The analysis of the gap, the formulation of goals and the way in which the goals should be reached is not something to be done by a specialist planning group or task force. The personnel concerned with the area in which the benchmarking is taking place must be deeply involved.

Install new methods

Implementation is the obverse of the coin of formulating goals and methods. If the development of goals has been well done and commitment obtained then a climate has been set for implementation. However, as with all the changes needed to gain world class manufacturing status, there needs to be a company-wide determination to achieve the benchmark standards.

■ 4.7 The product and the process

Throughout manufacturing industry there is a variety of design practices. In the consumer durables sector new products are launched at intervals depending upon the life cycle of the product. Life cycles are influenced by technological developments, fashion, competitors' actions, etc. The product may be developed on an incremental basis with innovations being incorporated into the product over an extended life cycle.

On the other hand the product may be totally new, a major breakthrough, where not only will the product undergo major innovation but so will the process. A major breakthrough in product development presents a much higher degree of risk. At the same time the company will have less experience of successful new product development than those companies pursuing a policy of incremental innovation. A policy of incremental innovation allows for the completion of the development cycle frequently leading to learning in how to manage product development and the allied process development.

From the quality viewpoint this relationship between product and process development is important. Designers have to evolve designs within the

Quality

capability of the process. This presupposes that the capability of the process is known and is under control. Neither supposition can be taken for granted.

Alternatively there may be reasons for including in the design components or features that are outside the current process capability which will result in the need for process development to cope with the new designs.

Figure 4.5 shows an integrated design approach identifying techniques and methods that can be applied to the stages of design and manufacture. The approach shows the integration required to achieve a manufacturable product that satisfies customer needs:

- Conceptual design fulfils the need identified by the marketing department.
- Detailed design provides the information required by manufacturing.
- It is also design that commits the organisation to the manufacturing cost.

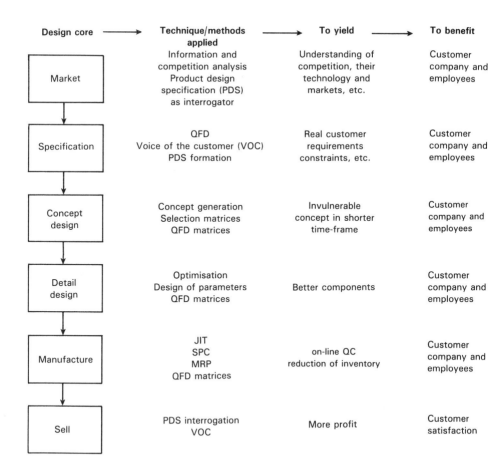

Figure 4.5 Integrated design (*Source*: Pugh (1991))

4.7.1 Cost and design

Although this chapter is concerned with quality it is important to recognise that design takes place within cost constraints. Tanaka (1989) identifies five stages in the development of a cost control system for new products, see Figure 4.6. He considers that the cost for a new product is based on target costing. The target cost is established so that the product will be profitable and will also be competitive in the market place. The target cost will be set low so that the design team is faced with a real challenge.

Target costs set for the product are subdivided into target costs for:

- Design.
- Manufacturing.
- Distribution.
- User activities.

Costs may be allocated on two bases:

1. Component blocks. Used for the design of new products that are similar to existing products. Generally such products will not incorporate new technology.
2. Functional areas. This is the choice for innovative product development. Functions are defined by use of matrices similarly to QFD.

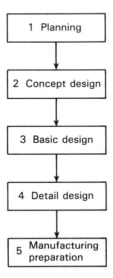

Figure 4.6 Stages in the development of a new product cost control system (*Source*: from *Japanese Management Accounting: A World Class Approach to Profit Management*, Yasuhiro Monden and Michiharu Sakurai (eds). English translation copyright © 1989 by Productivity Press Inc., PO Box 13390, Portland, Oregon, 503-235-0600. Reprinted by permission)

Quality

If, in the design phase, costs are not within target the design team must make changes to achieve target. Value engineering is the main technique used in Japanese companies to achieve cost targets.

4.7.2 Design time

An important consideration in product development is the time taken to get the product into manufacture. QFD, SE and other techniques for controlling new product development start the control once projects have been agreed. There is still time lost between identifying a potential or actual customer need and translating this into a design project. A considerable amount of time may already have been lost before a product design specification exists. This front end of design is a most difficult area to manage.

4.7.3 Product quality and process capability

Statistical process control (SPC) and SQC are techniques for improving the process and product quality. Readers are referred to the extensive literature on these techniques, details of which are outside the scope of this book.

These techniques bring about defect reduction as the process is examined and improved so that no more abnormal values occur. Nevertheless, as TSCo found, there is a limit that is reached in the reduction of defectives by the application of statistical control, as shown in Figure CS7.3, Case Study 7.

Statistical control will not achieve zero defects. It will reduce defect levels to the process capability. It will give the opportunity to effect process improvement. How to achieve zero defects will be considered in the next section.

4.7.4 Zero defects

Zero defects cannot be achieved by statistical methods and statistical methods are not applicable if the method of manufacture is moving towards JIT with a 'batch of one'. In that case it is not possible to build up statistical data, analyse the data and take corrective action – there is only one product.

Zero defects, right-first-time and JIT production are frequently associated. Leadtime control, reduction in inventory and responsiveness to customer demand is achieved by making small lots. Ideally these will be lots of one.

Right-first-time

If one-off products are made just in time it is imperative that these products should not be found to be defective at the end of the production process. Even

Quality circles

Illustration 4.5 Zero defects at TSCo

> An example of inspection to ensure zero defects at TSCo is in the manufacture of hi-lift springs. The outside diameter of the spring is obtained by hand grinding. This diameter is critical to the entry of the spring into its housing and is gauged after grinding. Occasionally customers return springs which will not enter the housing. This problem has been eliminated by putting a ring, of slightly smaller diameter than the housing, at the end of the grinding operation. Any springs that fail to pass through the ring receive further grinding.

if the defect is found part way through the process there may be problems in accessing machines to carry out rectification, and almost certainly the goal of smooth work flow, for other products, will be disrupted. Further, the product will not be produced just in time but just too late.

Elimination of defects is needed. Defects arise as a result of some condition which, if removed, would eliminate the defect entirely. Worker errors can produce defects, but only if the errors are not detected. Inspection to prevent defects has to be 100 per cent to be sure that no defects are passed on. Illustration 4.5, drawn from TSCo experience in their heavy spring division, shows the development of defect elimination by 100 per cent inspection.

This is an example of the application of autonomation, discussed in Chapter 2, Section 2.4.4. If the operation is machine controlled then it is necessary to apply defect elimination action to the set up of the machine. A method of checking the set up of the machine has to be developed so that each time an operation is performed a perfect unit is produced. It is not possible, as was the case with large lot production, to run off products, inspect and adjust the machine until perfect units are produced and then continue to produce perfect product.

The concept of autonomation outlined in Chapter 2 is vital to right-first-time. Everyone must be committed to stopping production while the reasons for the production of defective quality are eliminated.

■ 4.8 Quality circles

Quality circles (QC) or quality control circles (QCC) originated in Japan in the early 1960s. Up to that time the drive for quality originated by Deming and Juran had concentrated on SQC. This meant that the movement had been dominated by technicians with theoretical leanings. There was a realisation that the people who actually generated product quality were the shop-floor workers. As a result education and training began to shift to the workers. This gave rise to the quality circle movement.

There have been attempts to transfer QCs to America and Britain with

Quality

mixed success. One vital factor was that, in Japan, SQC had already made great inroads into problems of process capability and product quality. Thus there was already a quality climate and a record of success which workers were able to develop further by identifying problems and solving them. In Britain the development of SQC was not, in the majority of cases, well developed before QCs were introduced.

In Japan QCs were groups of employees meeting at regular intervals to discuss process problems. The leaders of the QC teams were not the supervisors. In attempting to transfer QCs to Europe/America not enough thought was given to the state of development of the quality movement and, also, to the different developments that had taken place in the management of the production system in Japan.

The value placed on QCs by the three countries is detailed in Table 4.6.

The low ranking given to past pay-off in both the US and the UK shows that Western culture has difficulty, at this stage, in accepting a concept founded on Japanese culture. The future emphasis in the US and UK continues to be low. In Japan, on the other hand, pay-off in the past has been high. As has already been commented, the major impact has already been made and future emphasis is almost certainly a reflection of the quality battle already won and the movement forward to flexibility and innovation.

There appears to be a movement in the UK to improvement teams rather than QCs. These operate under supervisors and are work teams. In TSCo, Case Study 7, Fisher, Case Study 2, and Kodak, Case Study 4, which are companies that have developed quality control to a level where the processes are under control, improvement teams are working.

TSCo considers that it achieved commitment of the workforce to quality by using a manual rather than a computerised SPC system. The manual system involved operatives understanding SPC and initiating corrective action.

Fisher found it gained the advantage of improvement teams from people empowerment. Kodak improvement teams focused on kanbans and in the process of improving kanbans effected improvements in quality.

The significant aspect in each case was that the workforce felt that they were involved. As a result they owned responsibility for quality and were prepared to work towards solving problems.

Table 4.6 The importance of quality circles

	Ranking of 26 programs and activities		
	US	UK	Japan
Pay-off in the past	19	21	4
Future emphasis	24	20	15

1 = most important 26 = least important
Source: Derived from *Benchmarking Global Manufacturing*, J.G. Miller, A. De Meyer and J. Nakane, Business One Irwin, 1992

Illustration 4.6 Quality circles at Nissan Manufacturing UK

Three things were needed to make QCs successful in NMUK:
1. The programme had to be about something more than quality.
2. It should be a natural extension of the way a team normally operated.
3. An external bureaucratic structure should not be set up which would frustrate achievement.

NMUK was committed to a policy of team working and development of individuals. QCs were seen as means of improving individual and team development and the participation of staff in the general day-to-day running of their working areas. QCs were to be fully integrated into the job and not seen as a separate activity.

A steering committee under the chairmanship of the director of production was set up to plan the more formal side.

Since problem solving and continuous improvement are part of everyone's job there was to be no financial reward for achievement. For the same reason no suggestions scheme was set up. Non-financial reward was important and was in the hands of the steering committee.

Source: P. Wickens, *The Road to Nissan*, Macmillan, 1987

The issue appears to be the way in which people are involved and the value the company puts on QCs or similar activities. Illustration 4.6 shows the Nissan Manufacturing UK (NMUK) approach.

■ 4.9 People empowerment

People empowerment will be discussed more fully in Chapter 5. The importance of people empowerment to quality is such that a brief exposition is warranted at this point.

People empowerment, in this case the ownership of quality, takes place as SQC, zero defects and QCs or improvement teams are introduced. Responsibility for the achievement of quality rests with the producer whether this is the designer producing a detailed design, the cost accountant producing control information or the operative producing product. Management must be aware, in this transfer of responsibility, of the change in culture that is entailed and the commitment and support that will have to be given in order to capitalise on changed relationships and methods of working.

As was seen in the Fisher Case Study people involvement was one strand of the manufacturing strategy. Subsequently, people empowerment became the driving force of the strategy. As ownership of quality is taken on by the shop floor so they want to make decisions.

The Times, 23 April 1993, comments on changes at Rover 'Rover has now taken the suggestion idea a step further, giving workers permission to sort out any problems they find in the business by forming "flying squads". The quality action teams, as they are known, are voluntary and a loose association of workers ready to pit their brains against a stumbling block in the efficiency of the business. In 1990, there were just 85 teams of eight people; last year there were 370 teams'.

■ 4.10 The control of supplier quality

There is a variety of supplies used by manufacturing industry. The quality of these supplies can be specified and controlled with varying degrees of precision. For example, in the case study companies minerals supplied to DSF Refractories can neither be specified nor controlled as accurately as mechanical components bought in by Stanley Tools. However, for that company bought-in castings will only reveal defects when machined, therefore consistent quality can only be obtained by the foundry developing excellence in their process.

Steel purchased by TSCo has to comply with a metallurgical specification and a dimensional specification. Although a sample can be taken from each coil of wire supplied, conformance to specification of the sample does not guarantee conformance throughout the coil. The only guarantee of conformance will come from the supplier's process being under control.

4.10.1 Supplier process capability

If the quality of product required is to be supplied to specification then that specification must be within the capability of the supplier's process. Process capability has already been discussed in Section 4.7. This raises three questions in choosing suppliers:

1. Does the supplier know the capability range of its manufacturing process?
2. Is the specification required within the capability range?
3. How does the supplier propose to monitor to ensure that product produced is always to specification?

4.10.2 Supplier appraisal and vendor rating

These questions can be answered as part of the appraisal process in choosing new suppliers. For key supplies visits will be made to potential suppliers to ascertain process capability together with systems of process and product quality control.

For items of lesser value information held by the purchasing function will help in the choice of supplier. BS 5750/ISO 9000 certification will be an indication of supplier commitment to quality procedures.

In conjunction with appraisal, companies need to monitor the performance of suppliers on a systematic basis. Vendor rating is a system for improving the performance of suppliers. For vendor rating to be effective, unsatisfactory suppliers must be given the opportunity of making improvements. The objective is to maintain high standards of supplier performance by regular feedback of performance.

4.10.3 Supplier partnering

Traditionally the Western relationship with suppliers has been adversarial. For key supplies more than one supplier has been used, in part to obtain security of supply, but generally to retain a choice. The choice of where to place orders has been made on the basis of price. Suppliers who meet quality and delivery criteria are monitored to identify which offers the best price at the time that an order is to be placed.

The Japanese have built a system of supplier partnering where a long-term engagement has been made with a supplier. Both sides of the partnership have expectations of gain. The supplier, by virtue of a stable relationship, can plan capacity. Process development can be undertaken with confidence that future orders will be placed which will allow the supplier to recoup the cost of the development. Partnership in design has allowed the supplier to utilise its expertise to improve products and methods of production.

The customer expects the gain of consistent quality to give savings in inwards goods inspection as well as a continuously reducing price of product. This is expected to be achieved by the supplier because the stability of demand allows improvements in the process of manufacture. These improvements are expected, not only to reduce the price to the customer, but also to increase the supplier's profits. In this way there is gain to both supplier and customer in supplier partnering

Evidently for such a partnership to operate trust is the critical factor. As with the development of quality in Japanese manufacture which by the 1990s had extended over a period of some 40 years, so supplier partnering in Japanese manufacture has a long history. Additionally the Japanese system of trading groups, where customers have interlocking investments in their suppliers, promotes supplier partnering.

Supplier partnering is being developed in the West. Whether the climate of trust exists which will bring about the change in the Western long-term adversarial relationship remains to be seen.

Practically, supplier partnering requires reduction of the supplier base. Fisher Controls, for example, has reduced the supplier base from 560 to 154 over a period of four years. Nevertheless, industry comparisons show that the number of suppliers used by Western manufacturers exceeds the figure for Japanese manufacturers by a factor of between 3 and 6.

■ 4.11 Continuous improvement

Referring to Table 4.1 it can be seen that continuous improvement is a common theme in the four approaches to quality improvement. It is salutary for managers to consider the implications of continuous improvement:

- There needs to be a commitment throughout the company to work towards improvement of all aspects of the business; systems, methods, processes and products. To obtain such improvements the capability of the personnel of the company must be improved by programmes of training and development. A culture has to be fostered in the company that will encourage improvement. The company has to make a large investment to obtain the benefits of continuous improvement.
- Improvement has to be made so that the company gains competitive advantage. While all aspects of the business may be capable of improvement management must direct the improvement effort into those areas that are going to give the best payback.
- Continuous improvement results in increased productivity. Improvement in the use of resources results in more being produced from the existing resource or, alternatively, less resource being needed. Unless increase in sales can be achieved to match the level of improvement made, there will be redundant resource. When this is human resource managers have a difficult task to maintain a programme of continuous improvement which will result in redundancy.

■ 4.12 Conclusions

Cost of quality has a considerable influence on how quality is managed. Once there is an acceptance that the cost of quality is many times higher than the cost identified by traditional costing systems, then management effort to reduce those costs becomes worthwhile.

As processes are brought under control it is recognised that costs are committed at the design stage. It is at this stage that cost, and therefore profit, will be determined. The opportunity to effect cost-saving improvements during production are limited particularly where short product life cycles lead to frequent design changes. The interface between design manufacture and materials supply must be carefully managed. It is essential to set quality parameters which satisfy the customer and which can be achieved by the company and its supply chain.

Conclusions

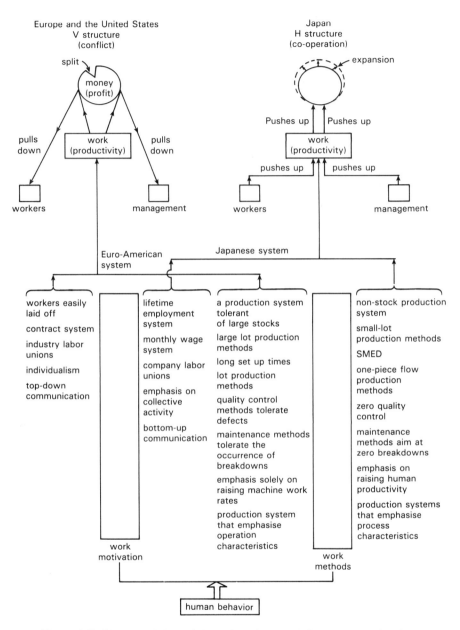

Figure 4.7 Characteristics of Euro-American and Japanese production systems (*Source*: from *Zero Quality Control* by Shigeo Shingo. English translation copyright © 1986 by Productivity Press Inc., PO Box 13390, Portland, Oregon, 503-235-0600. Reproduced by permission)

123

Great importance is attached to TQM in the effort to reach world class manufacturing status. As this chapter has shown, there are numerous ways in which manufacturing management can attempt to reach the goal of TQM.

There is no uniquely correct way of achieving TQM. The needs of manufacturing companies vary depending upon their markets, products and processes. Each company must make a choice in the light of its own circumstances, and recognise that it is setting out on a long and arduous road where determination and staying power will be a substantial factor in success.

Benchmarking is a technique which offers the possibility of improving products, processes and systems and, therefore, total quality. Continuous improvement is the goal. The way to reach this goal is seen to be the empowerment of people to control and improve quality.

However, Shingo's (1986b) analysis of the characteristics of Euro-American and Japanese production systems shows the major changes which need to be made and suggests that a gulf exists between Japanese and Western approaches, see Figure 4.7.

The main differences lie in three areas:

1. Culture, typically Western individualism versus Japanese collective activity.
2. Production systems, Western long set-ups, toleration of defects versus Japanese SMED and zero defects.
3. Japanese worker/management push versus Western worker/management pull.

As with all attempts at comparison exceptions can be found where Japanese characteristics are inferior to Western characteristics. Nevertheless the figure provides a thought provoking framework for the achievement of zero defects. While changes are being made in the Euro-American system there is no doubt that the characteristics that have achieved the very high level of quality in Japanese manufactured products have not been totally adopted in Western manufacturing.

5 The Management of Human Resources

■ 5.1 Introduction

Figure 5.1 shows the interaction between unions, workers and management that is needed to achieve competitive manufacturing. Competitive manufacturing requires a strategic approach to the development and satisfaction of markets that will enable a company to compete with the best in its industry.

Those companies which only market in Britain are competing with foreign products which their products must equal or surpass. The knitwear industry is an example where many companies supplied only the home market. Large numbers have gone out of business in the face of foreign competition.

A critical factor that enables a company to achieve competitive manufacturing is the human resource. Attempts to satisfy customer needs by adopting practices such as TQM and JIT, or to improve the manufacturing process by the application of computer controlled processes and methods will be frustrated if the human resource is unable, or unwilling, to support these changes.

A problem in considering the human resource is contained in Figure 5.1 where management and workers are shown separately. The human resource of the company comprises all the personnel working for that company. The goals of competitive manufacturing demand that the combined human resource of the company is committed to the improvement of the company's competitive position. Management has to take the initiative but must be aware that workers have a significant contribution to make. This contribution can be encouraged or stifled by management.

The way in which the human resource is managed has evolved during the long history of industrial development in Britain. As Chapter 2 shows manufacturing contribution to the GNP over the period 1970 to 1990 declined faster in Britain than in other industrialised nations. However, that industrial decline had its origins much earlier in time. The decline in manufacturing stems from many causes but inefficient use of the human resource has been an important factor. This chapter will examine ways in which human resources

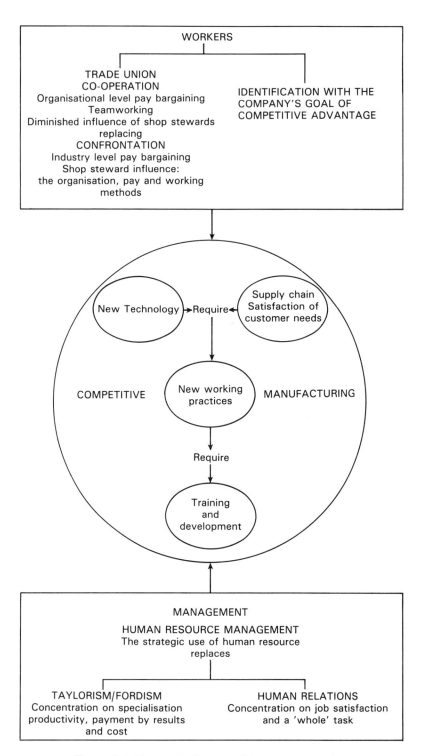

Figure 5.1 The goal of competitive manufacturing

The management of the human resource

have not been used effectively in manufacturing and will examine ways in which those resources can be used more effectively. Throughout the discussion it must be remembered that ineffective use of the human resource is a problem of the way in which it is managed. How human resources can be managed effectively in the future depends on the initiatives taken by management.

■ 5.2 The management of the human resource during the twentieth century

5.2.1 Scientific management

The use of scientific management led to workers being regarded as 'extensions to machines'. Specialisation resulted in workers being responsible for small tasks which they had difficulty in relating to the main activity of the company. Work was boring and repetitive and money was the only motivating factor. Strikes both official and unofficial were frequent during the period from the 1950s to the 1980s. Figure 5.2 shows the pattern for manufacturing disputes between 1960 and 1992.

Disputes arose over the rates to be paid for new work when it was introduced and also over management attempts to restudy work as the 'best method' became outdated due to learning curve effects. Multiple rates arose for the different skills involved. Good paying jobs were sought by operatives

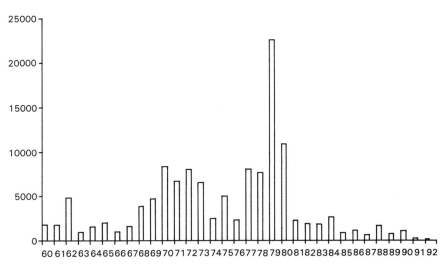

Figure 5.2 Days lost in manufacturing disputes (000s) (*Source*: adapted from data published by the Central Statistical Office)

127

The Management of Human Resources

while bad paying jobs were neglected. Shop stewards assumed a significant role in the negotiation of rates and this local level bargaining became more important than main wage negotiations at national level. Personnel managers were needed in negotiations and in the settlement of disputes. Wage drift was common. Wage drift is the way in which the level of pay received by the operative drifts upwards away from the level based on pre-determined levels of effort set by work study.

5.2.2 Human relations

Another approach to the management of the human resource arose which was labelled human relations management. Once more the origination was in America where the Hawthorne Investigations took place in the 1930s. These investigations were intended to identify the effect on productivity of changing environmental conditions. The environmental conditions for an experimental group of workers were varied. Output increased as the conditions were changed. Even when environmental conditions reverted to those obtaining at the start of the experiment output continued to rise. The conclusion was reached that the involvement of the group and respect for them by the management were critical to the increase in output.

The application of human relations as a way of managing the human resource became general after the Second World War but only as an overlay on scientific management. During this period the work of Maslow and Herzberg led to a questioning of the role of incentives as the motivational factor. During the 1970s the level of industrial disruption due to strikes led to a questioning of the PBR system and the nature of much production work. There was an effort to build up the work content so that workers had a bigger task. Job enlargement, job enrichment and job rotation were practised. An attempt was being made to involve workforces just as the experimental group had been involved at the Hawthorne plant.

5.2.3 Managing the managerial resource

The previous sections outline management's attempt to manage the human resource at the shop-floor level. MBO was an attempt by management to manage the human resource of the company at the managerial level, i.e. management's attempt to manage itself. The general failure of MBO as an effective way of improving performance has already been outlined in Chapter 3, on p. 87. Nevertheless, MBO was an attempt at promoting greater management involvement in the company and its objectives. Managers, like shop-floor workers, performed specialised tasks. They were encapsulated in functions remote from the environment and the customer.

5.2.4 Failure in managing the human resource

For much of the twentieth century British manufacturing has failed in its management of the human resource. Criticism must be made on three counts:

1. The application of Taylor's scientific management continued in Britain over many years with little development. A 'them and us' syndrome arose in British industry due to the separation of planning and doing. This was reinforced by status divisions brought about by different conditions of employment, starting time, canteen areas, parking, etc.

   The shop floor was paid by results so that earnings depended on performance. Managers received their salary whatever the level of performance, thus increasing the feeling of 'them and us'.

   This unchanged application of scientific management continued even though there was the example of American companies moving into the UK with quite different working practices, in spite of their use of scientific management.
2. Management allowed PBR and its negotiation to usurp the authority of management to run the business.
3. British management adopted techniques involving the human resource and subsequently abandoned those techniques. MBO has already been instanced but value analysis/engineering, and quality circles have been tried and, largely, abandoned. MRP has had more failures than successes.

This raises a question for the future as to whether manufacturing management has the will and tenacity to make a success of the changes that will come with the adoption of human resource management.

5.2.5 Human resource management (HRM)

HRM is a term that is currently applied to the strategic use of human resources. It describes the range of policies that are strategically significant in the use of human resources such as:

- Organisational integration.
- Employee commitment.
- Flexibility.
- The quality of work.
- Productivity.
- Changes in organisational values.

Case Study Extracts 5.1 and 5.2 show ways in which HRM can be used strategically and the benefits that have been obtained.

The Management of Human Resources

Case Study Extract 5.1 Fisher Controls (Case 2)

> Fisher had to formulate and implement a manufacturing strategy if it was to survive. Competitors' performance was known to be much higher and if Fisher's manufacturing performance was not improved manufacture could be sub-contracted. Although the Leicester site was not greenfield the site had only been occupied for three years. There was no union resistance.
>
> The manufacturing strategy addressed five areas:
>
> 1. People involvement.
> 2. Total quality
> 3. Supplier partnering
> 4. MRP/JIT
> 5. Loss prevention.
>
> There was extensive consultation with the workforce about where the company should be in two to five years time although the evolution of the strategy was firmly in the hands of the management.
>
> The strategy that was developed and how it was to be implemented was communicated to manufacturing personnel via the management supervisory structure. Detailed strategies were developed for each of the five areas.
>
> To achieve the goals of the manufacturing strategy production was reorganised to a product line structure of cells and teams. Teamworking brought about people involvement and ultimately people empowerment which was necessary to achieve the goals of the manufacturing strategy in the areas of TQM, MRP/JIT and loss prevention.
>
> The improvement in performance due to this change is shown in Figure CS2.4.

In the case of Fisher Controls the HRM strategy supported the manufacturing strategy. The strategies were not planned at the start of the period. There was a vision, and detail was developed during the period 1988–1992 as successes were achieved and new goals could be set.

The Kodak MRP introduction is part of a worldwide improvement process initiated by Kodak Eastman. The case study extract demonstrates the commitment needed by all company personnel to achieve the goals of a strategy such as the introduction of MRPII and the benefits that result from such commitment.

These case study extracts demonstrate the strategic nature of HRM. The initiative for the strategic decision comes from management with the involvement of the shop floor. Implementation is a combined management/workforce activity flowing from the strategic decision. Both management and shop-floor workers have to learn and adapt to the changes brought about by the strategy.

The best Japanese companies have used human resources as an integral part of their strategy, in manufacturing companies. The human resource

The management of the human resource

Case Study Extract 5.2   Kodak (Case 4)

> **MRPII introduction**
>
> An aggressive schedule was adopted of between 18 months and 2 years. The short time scale was adopted to:
>
> - Generate intensity and enthusiasm.
> - Focus on MRP/kanban as the priority.
> - Regard MRP/kanban as the vehicle driving the changes needed to achieve success.
> - Prevent schedule slippage.
> - Realise benefits quickly.
>
> In December 1988 a steering team was formed which agreed to:
>
> - Adopt the Oliver Wight Associates' 'proven path' as the implementation approach.
> - Form a commissioning team.
> - Provide an education programme for the management team.
>
> Education and training were given as the need arose. Posts were created and, in one case, an unsuitable manager was replaced. Within six months a pilot introduction had been made and production and operations managers were owning the decisions regarding planning, material and information flows within their areas.
>
> Twenty-two months after starting introduction Class A status was achieved. At the peak of the introduction 6 per cent of the workforce was involved in the MRP activity while production and satisfaction of customer orders continued. Education and training were needed throughout the introduction.
>
> The benefits achieved from the successful implementation of MRPII are:
>
> - Manufacturing leadtime reduced by 70 per cent.
> - Inventory reduction 40 per cent.
> - Storage space reduction by 80 per cent.
> - Item level performance in excess of 95 per cent.
> - Labour productivity increased by 40 per cent.

strategy has been particularly significant in the automotive industry and has enabled Japanese companies to gain a competitive advantage that has given them a lead over automotive companies in the US, UK and the rest of Europe.

Monden (1989a) in examining the Toyota production system identifies the most important goal as the reduction of costs. This emphasis on cost reduction appears to contradict the criticism that has been made throughout this book that Western management concentrates on low cost. The difference between Western and Japanese companies is that the former concentrate on an efficiency approach to reducing the costs of production while the latter concentrate on the elimination of waste. The Toyota waste elimination philosophy has been discussed in Chapter 2, p. 15. Japanese manufacturing companies setting up operations in the UK have successfully managed British workforces to achieve levels of performance equalling those of the parent companies in Japan.

The current emphasis on HRM by British manufacturing management is an attempt to catch up with the Japanese in the strategic use of human resources. As already described, British manufacturing management, particularly in the large mass production companies, has grown accustomed to the use of the principles of scientific management in their attitude towards the shop-floor workforce. Some changes have been made but management is slow to move to an integrated workforce where everyone shares the responsibility for continuous improvement.

Effective use of human resources needs commitment by all employees of the company. As technology is introduced the human resource employed in production operations can be reduced in numbers. To be effective in operating production lines and machines with reduced manning levels, functional flexibility and skill levels have to increase. The organisation at the workplace and in the rest of the company changes. These changes have to be made within a framework of industrial relations (IR) which has been built up over decades to protect worker interests under old methods of manufacture.

■ 5.3 Unions, workers and management in competitive manufacturing

5.3.1 The changing role of the trade unions

The 1950s saw the rise of nationalised industries, full employment and the introduction of the welfare state. Manufactured goods were in short supply throughout the world. Large areas devastated by the war needed rebuilding and it was a seller's market where the emphasis was on manufacturing since everything that could be made could be sold.

By the 1960s trade unions were strong and had adopted a system of industry level bargaining. Such bargaining provided only basic terms of employment. At plant level actual wages depended on schemes of PBR that were used to obtain levels of productivity. Management was generally weak and reactive to union pressure. Shop stewards were influential in the bargaining process and also in decisions about manning levels and, therefore, productivity. Strikes were a powerful weapon used unofficially by shop stewards and officially by the unions. It can be seen from Figure 5.2 that during the late 1960s and the 1970s industrial disputes resulted in considerable disruption to work. This scenario of industrial conflict was the background to government attempts to reach a relationship with the unions that would result in improved relations between employers and unions.

The Donovan Commission (1968), set up by the government to investigate the conflict between workplace and industry level bargaining, recommended a voluntary reform of industrial relations with management taking the initiative. The government, disappointed by the voluntary nature of the recommendations published its own White Paper *In Place of Strife* (Department of Employment and Productivity, 1969). In the event the government backed off from implementing legislation that would have led to a confrontation with the unions.

The 1970s saw a further rise in the power of the trade unions. Employment legislation gave employees and unions a range of rights which strengthened the position of workers and unions and as a consequence weakened the authority of management. The closed shop assumed increasing importance in employer/worker relations particularly in manufacturing industry.

Unions were increasingly involved with the government in the economic policy of the country. The government attempted to control inflation by involving unions in wage restraints. As a result inflation that had reached a high of 24 per cent in 1975 fell to 8 per cent in 1978, but by this time the unions were no longer willing to observe wage restraint. The 'winter of discontent' followed where major strikes, particularly among public sector employees led to widespread disruption of services throughout the country, and ultimately, to the fall of the Labour government.

In the 1980s there was a succession of Conservative governments under Margaret Thatcher dedicated to taming the power of the unions. Five major pieces of legislation were enacted:

1. Employment Act, 1980.
2. Employment Act, 1982.
3. Trade Union Act, 1984.
4. Employment Act, 1988.
5. Employment Act, 1990.

The main effects of these Acts were:

- A weakened closed shop. This resulted in employers dealing directly with individual workers not indirectly through a shop steward.
- Secret balloting became the norm for the conduct of unions, whether for contributing to political funds, the election of officials or taking strike action.
- Secondary action was rendered illegal.
- The immunity of trade unions and officials was removed or reduced. They were made responsible for unofficial strikes. The effect of this part of the legislation was that unions could be pursued through the courts for their actions.

By the 1990s the unions were in a much weaker position due to government legislation. In manufacturing the power of the unions had been further weakened as the manufacturing base declined in importance and the drive for

The Management of Human Resources

productivity to counter foreign competition led to huge job losses. Figure 5.2 shows the decline, from the mid-1980s, in days lost from manufacturing disputes.

International competition had eroded large areas of manufacturing. An early casualty was shipbuilding which had been unable to effect the change from its traditional methods of manufacturing rapidly enough to compete with Japanese and Korean methods of shipbuilding. The durables sector saw whole industries collapse such as motorcycles, television and hi-fi. The automotive industry was under considerable threat from imported models. Shipbuilding in the UK has continued to decline, but the durable sector industries such as television, motorcycles and automotive are now in a strong competitive position, due in part to Japanese inward investment and to the adoption of new manufacturing philosophies.

The rise in the import of Japanese goods in these markets slowed as Japanese companies set up manufacturing operations in the UK. This move was stimulated by the development of the European Single Market and the need for Japanese manufacturers to establish bases from which they could exploit it. This was particularly notable in the car industry and led to the establishment of manufacturing operations by Nissan, Honda and Toyota.

The outcome of the introduction of new technology and the resultant increases in productivity have caused a continuing decline in the level of employment in manufacturing as can be seen in Figure 5.3. Some 3.5 million jobs have been lost in manufacturing between 1974 and 1993.

The effect on the unions of legislation together with the high level of unemployment through the 1980s and early 1990s brought about a decrease in trade union membership of some three million during the 1980s, see Figure 5.4.

This decline in trade union membership is not due solely to the decline in

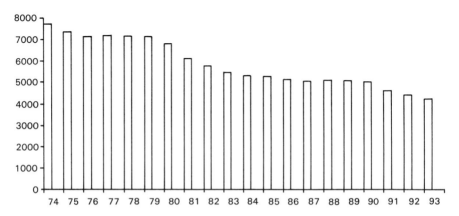

Figure 5.3 Employment in manufacturing (000s) (*Source*: adapted from data published by the Central Statistical Office)

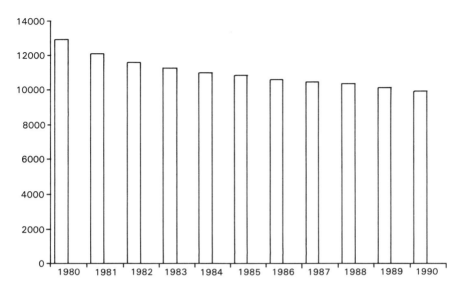

Figure 5.4 Trade union membership in the UK (000s) (*Source*: adapted from data published by the Central Statistical Office)

numbers employed. Union density, which is defined as total union membership expressed as a percentage of the civilian workforce in employment, has declined during the 1980s, as shown in Figure 5.5.

5.3.2 Changes in trade union and workforce response to new technology

During the 1980s there is some evidence that management moved towards managing the human resource strategically. The introduction of HRM has been a response to increased competition. Undoubtedly this response has been influenced by Japanese practice. Many companies realised that the advantages to be gained by introducing new technology could only be realised if the human resource made effective use of the new technology. Companies began to realise that considerable improvement could be made in the use of existing technology by improving the effectiveness of the human resource.

Japanese inward investment succeeded in establishing viable companies in sectors where British manufacturing had been eliminated or is in decline, particularly in setting up plant for automotive manufacture. The fact that the Japanese, using a British workforce, could set up operations that were competitive with the best in the world had an influence both on methods of manufacture and the use of human resources in British companies. British manufacturing companies started to raise their levels of quality and productivity towards those which the Japanese had demonstrated could be obtained by the manufacturing companies which they had set up in the UK.

The Management of Human Resources

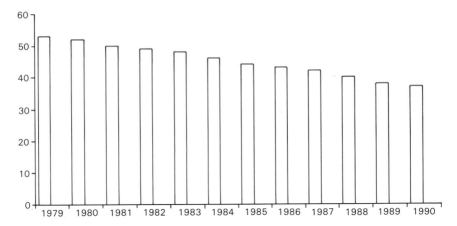

Figure 5.5 Union density (*Source*: adapted from data published in the *Employment Gazette* (various years))

A major factor in the improvement of manufacturing has been the introduction of new technology in conjunction with new methods of working. This requires changes in skills and organisational changes. The union response in Britain has favoured the introduction of new technology and 'new technology agreements' but these agreements are often procedural rather than effective negotiation on the introduction of new technology. Unions are willing to co-operate in the introduction of new technology but are resistant to organisational changes in methods of working.

These changes involve team working based on manufacturing cells with worker empowerment. Such changes approach the Toyota concept of autonomation (the control of defects by the production line). These methods of working leave little room for the traditional role of the union. The unions do not want to see their power base eroded and would rather have participation in which the traditional bargaining roles are preserved.

5.3.3 What is needed to manufacture competitively

Milsome (1993) is critical of the generally accepted view that adoption of Japanese manufacturing practices will lead to success in British manufacturing industry. The article considers, instead, that Japanese competition is a contributing factor to the adoption of a range of new manufacturing methods and working practices in British manufacturing.

The author judges that the new manufacturing methods and working practices shown in Figure 5.6 are needed to raise the competitive level of British manufacturing.

The argument made throughout this book is that by integrating the

Unions, workers and management

activities of the manufacturing company and focusing them on customer satisfaction, the competitive edge of a company will be improved.

The adoption of manufacturing methods and working practices must be company specific and involves consideration of the four activities shown in Figure 5.6, which is a development of 'competitive manufacturing' in Figure 5.1.

1. New technology. This is the application of technology to processes and operations as well as systems and methods of control. New technology can lead to integration and it can lead to flexibility. How well either will be achieved depends on the analysis of manufacturing problems and opportunities in the individual company.

   The introduction of new technology has to be effectively implemented. The benefits have to exceed the costs. To implement

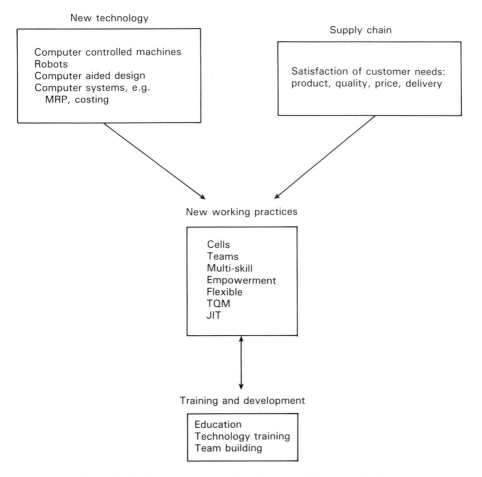

Figure 5.6 Human resources and competitive manufacture

effectively and realise benefits in excess of cost are challenges for management.
2. Supply chain management. The aspect of integration in which the customer is the focus has been examined in Chapter 3, Section 3.5.2. The ability of the company to satisfy the customers' needs better than the competition will give the company advantage over these competitors. The company depends on a network of external resources in the manufacture of products. These resources, in combination with company resources, have to be organised so that the goal of customer satisfaction can be achieved. These external resources are not owned by the company and have to be managed so that mutual profit results. HRM strategy in these supply chain companies becomes important to the manufacturing company. Influencing HRM strategies in other companies is difficult but, if acceptable leadtime and quality are to be obtained, the supply companies and manufacturing companies need to develop such relationships.
3. New working practices. These are needed to exploit new technology successfully, to realise effective supply chain management and to exploit its potential. These new working practices shown in Figure 5.6 are readily listed, easily comprehended and very difficult to realise. Their effective introduction needs leadership, training and development of a very high order.

    Each company needs to pursue its own strategy for new working practices and to build its own culture. To gain competitive advantage the company must be able to do some things better than its competitors can.
4. Training and development are vital if a company is to be successful in implementing new working practices. Section 5.4 will analyse the problems in this area and the actions being taken.

Human resource management

In attempting to become more competitive manufacturing companies are altering the way in which they manage and use human resources. Management now, in contrast to the practice under scientific management, wants a flexible workforce, able to think creatively and be 'empowered', to be responsible for their output and standards of quality. The workforce is expected to work in a supplier/customer relationship with the next stage of manufacture. Management wishes to extend worker responsibility to include maintenance of a routine nature and also to work with maintenance teams when breakdown occurs. Maintenance personnel in turn are required to be multi-skilled.

These management initiatives are bringing about a workforce with different skills which is expected to assume many activities that, previously, would have been undertaken by management. Trade unions need to examine their own goals and the ways in which they represent their membership. They have to make a response to this new management initiative. It is difficult for the unions to have a co-ordinated response when several unions represent specific groups of workers within a company rather than one union representing all company employees. Representation in the past has mainly been concerned with conditions of employment. Unions faced with a strategic approach by management in the way the human resource should be used are in a learning situation in making their response.

The TUC has attempted to make a co-ordinated response to HRM in three ways:

1. The need for the approach to be realistic. A new realism arose in the trade union movement in the 1980s that considered compromise and partnership to be the best practice. This was an endeavour to make a positive contribution within a changing manufacturing environment.
2. An emphasis on the role of collective bargaining in employee development, participation and equal opportunities. This emphasis is on modifying existing practices to respond to the changes taking place.
3. The recognition that there are elements of HRM practice which undermine the independence of trade unions, e.g. company-oriented forms of worker representation and team structure. These displace trade union representatives as the traditional form of communication and representation. The result of this recognition is a resistance to change and a desire to retain the traditional relations and work practices.

The 'New Agenda for Bargaining in the 1990s' was adopted by the TUC in 1990. There is an emphasis on collective bargaining and a support for co-operation rather than confrontation with management. There is reference to the extension of union involvement in issues such as training and participation.

HRM has brought about a reassertion of corporate identity and behaviour with the company as the focus of IR activities. Companies have decentralised management structures resulting in decentralisation of collective bargaining. The decentralisation can be seen internally through the introduction of SBUs and cost centres, externally through sub-contracting or revenue sharing partnerships. Japanese companies consider enterprise level arrangements as fundamental to IR systems and these ideas have gained acceptance in the UK with the rise in Japanese inward investment. Thus the system of industry agreements that was a feature of manufacturing industry in the 1960s and 1970s is being replaced with systems of company agreements that are often devolved to plant level.

5.3.4 Management implementation of new methods to raise the competitive edge

Criticism has been made in this book of the way in which management applied scientific management. Over many decades there was little regard for changes

The Management of Human Resources

in the environment which should have resulted in change in methods of managing.

HRM has brought about a reactivation of management which has become assertive in demanding greater employee commitment. Workers are becoming empowered to control quality. In cell manufacture the team is responsible for scheduling production within the cell. Internal customer relations are introduced where each successive operation is regarded as a customer of the previous operation and workers are given responsibility for the work being carried out. There is an increase in the demands being made on workers and a criticism of the practices is examined on p. 141.

All this is a far cry from the separation of planning and doing that was the basis of Taylor's scientific management. Manufacturing to a pull system of customer demand is equally far removed from the Ford assembly line. Nevertheless, when commitment is closely examined it can be seen that the resultant empowerment controls the worker as closely, or perhaps, even more closely than the Taylor/Ford system it is replacing.

Management must be aware of the changed demands that new systems and methods of working make on the workforce. These demands must be carefully managed if the new style of management is to succeed and continue to succeed. Commitment by the workforce resulting in improved quality delivery and cost will lead to an expectation of increased reward. The method of reward needs careful examination if commitment is to continue. The need to make sure that rewards remain aligned with performance must be addressed by management.

On greenfield sites companies have been highly selective in recruiting their workforces. Recruitment has not necessarily been from workers skilled in the industry. The adoption of JIT/TQM systems on greenfield sites has been made easier by the negotiation of single union agreements, frequently with no strike agreements.

The adoption of such systems on brownfield sites is much more difficult. The reasons for brownfield companies having more difficulty than greenfield companies in adopting JIT/TQM systems and making other changes that are needed to gain competitive advantage are not difficult to assess. The workforce exists and has been trained to work under existing methods of production. The union relationship is already established and the culture of the company has developed since its establishment. The author is optimistic that progress can be made. The case studies demonstrate that some companies have effected considerable changes. It is management's task to analyse their particular circumstances and develop an HRM strategy. Management must take an initiative in involving the unions in implementing the HRM strategy. Changing brownfield companies where the change is to the culture of the company needs dedication and determination. This is only possible if management believes in what it is trying to do.

Case Study Extract 5.3 demonstrates some of the difficulties faced by management in introducing new technology.

Case Study Extract 5.3 Introduction of the auto-line at TSCo (Case 7)

> This line had been under consideration for some five years. The managing director had visited Japan in 1980 to see an automated line. However, the decision had been delayed in part due to the cost of investment but also due to organisational and workforce considerations.
>
> During 1982 TSCo was reorganised into five divisions. One of these was the valve spring division which allowed the newly appointed divisional manager to prepare a specification and layout for an auto-line.
>
> It was important to have a team approach to the introduction of an auto-line. In 1982 the union continued to adopt a militant confrontational stance. Relations between management and unions deteriorated. A point at issue was the introduction of new manning levels required by the reduced demand for company products. Disagreement resulted in a strike.
>
> The resolution of the strike gave rise to a much improved atmosphere which, it was felt, would give the co-operation and teamwork needed for the introduction of the auto-line.

Criticism of the new working practices

There is a body of academic writers who are critical of the JIT/TQM system. They regard the system of worker empowerment/ownership as one of worker exploitation. Workers are given additional responsibilities as if they were manufacturers in their own right, without the freedom of real manufacturers to sell outside the factory system. The system becomes one of total management control. Workers are more tightly controlled than under the Taylor/Ford system. By the elimination of waste they have to work harder.

This is not the view taken by John Edmonds (1990): 'we are conceding the enhancement of our members' jobs, the creation of more satisfying work, the creation of more fulfilling work, better opportunities, promotional opportunities, because they were all a disguise that the employer is using to grind us down. We cannot look backwards like that. We have got to seize some of this ground ourselves. Job enhancement should be ours, job flexibility should be ours'.

At present the fear is of redundancy and employees are interested in their companies' ability to compete. Consequently there is little resistance to JIT/TQM. In many companies the introduction of JIT/TQM has not yet happened. For some the Toyota/lean production system never will be appropriate. Nevertheless, management must be aware that JIT/TQM systems do make demands on workers. The advantages of the systems may be seen to outweigh the disadvantages but management must not make the mistake of failing to recognise how systems have to be developed to remain aligned with a changing environment.

The Management of Human Resources

■ 5.4 Developing the human resource

In a majority of factories the human resource required is a multi-skilled highly flexible workforce. The flexibility involves free movements between tasks. All levels within the organisation have to be committed to company goals which demands high levels of trust. Leaders are needed who are able to develop such a workforce that can respond to the changing needs of the company.

The development of the human resource needs:

- An ability and willingness to adapt to change and in the process to solve problems and make the change work.
- An ability and willingness to accept delegated decision making – people empowerment.
- An ability and willingness to accept training and to expect to have to replace existing skills with new skills during a working lifetime.
- An ability and willingness to be a part of an environment where continuous improvement is a way of life. This means that however good a process and its methods of operation may be today improvement is expected next day and the day after.

Working towards these goals demands positive leadership. Management must be committed to the development of the human resource. This commitment has to be able to respond to the ways in which process and product development affect the workforce. When problems arise management must be prepared to work with the workforce in solving the problems. The workforce is being asked to change from the Taylor system of specialisation with management as the planners to a system where there is ownership of activities across functional boundaries. Strong management support is needed particularly in maintaining the company strategy when things are not going well. On brownfield sites the implementation of a HRM strategy is complicated by the need to satisfy customer demands, while, at the same time, achieving the profit goals needed by the company to satisfy the providers of capital.

Successful HRM depends on:

- Education, training and development.
- Appraisal and reward.
- Flexibility.
- Leadership.

5.4.1 Education, training and development

Education, training and development overlap.

Education

The interest of manufacturing companies in education is whether there is a sufficient number of adequately educated entrants available. There have been numbers of reports over the years that conclude that the educational level attained in Britain is lower than in comparable industrialised countries.

The *Financial Times*, 26 March 1993, summarised a report from the NIESR:

- Only half the 16-year-olds in Britain attained an equivalent standard to those in France, Germany and Japan in science, native language and mathematics.
- The proportion of British 16–19-year-olds in full-time education and training was lower than in competitor countries.
- Fewer British young people aged 18-plus gained school or vocational qualifications than in France and Germany.

Earlier studies by NIESR have shown that the shortage of craft skills is the principal deficiency of British industry when compared with European competitors.

Training and development

Training has seen a series of initiatives with training boards being the vehicle used during the 1960s. Subsequently these were discontinued and the Manpower Services Commission became the government training agency. Training was allied to steps to retrain the unemployed and to provide a means of entry into work for school leavers.

In *The Times*, 22 November 1992, Gillian Shephard, Employment Secretary described current education and training at government level as 'a muddle'. However, she is optimistic that the skill-based awards, National Vocational Qualifications (NVQs), the Investors in People scheme (IiP) and the Training and Enterprise Councils (TECs) will give the level of training needed for Britain to become competitive.

In the same edition of *The Times*, Noriko Hama of Mitsubishi Research, dismisses British training 'Japanese businesses had to start from scratch to teach their workers to revere the consumer. The Japanese found it difficult to convince their British workers that the consumer was God and should be offered a fragrant offering. The British are very much happy to settle for second best both in terms of skills and what the consumer expects'. She considers that Japan and Germany remain at the pinnacle of training unlike Gillian Shephard who thinks that there is a greater enthusiasm in Britain than in Germany for workplace training.

A third article in the same edition deals with a nurse trained as a Toyota worker: 'She was sent to work on Toyota assembly lines in Japan and the

United States to learn every facet of car assembly.' These are examples of companies providing training in British manufacturing industry.

A similar approach to training and development is reported for Rover which has 10,000 of its 34,000 employees training at any one time at a cost of £120 million per year.

Management Today, December 1992, carried a five-article survey on training. It quoted the CBI and TEC initiative of the IiP scheme launched in 1980 having 77 firms by mid-1992. The target given by Gillian Shephard is 6,000 companies by 1996. Her own figure in *The Times*, 22 November 1992, is 106 companies which leaves a further 5,894 companies to be recruited in four years.

Other statistics and research in the article fail to inspire confidence. According to Professor Prais of the National Institute for Economic and Social Research there are reservations about the quality of NVQs. Companies are responsible for assessing their own youth trainees and get a financial bonus from the government for passing them.

A Cranfield/3i survey is quoted which finds that small and medium-sized British enterprises spend about 1/2 per cent less of their turnover on training than other EU countries and such training is biased towards already qualified personnel.

An appreciation of the size of the problem facing British manufacturing companies is that if German standards of qualification in engineering are to be matched in Britain then an extra 80,000 people a year need to be trained to BTEC National Certificate or City and Guilds standards.

Glynn and Gospel (1993) state that 'British workers are on average less well educated and trained than those in other advanced industrial nations'. As a consequence British industry produces less sophisticated products which have lower unit values, use simpler processes and less skilled labour. This leads to 'a low skill equilibrium' which presents problems in raising the level of technology in use in UK manufacturing.

Development is an on-the-job activity that takes place in a variety of ways.

- Quality circles and improvement teams are forms of employee development.
- The introduction of techniques such as MRP or SPC involves some training plus a high degree of employee development that will continue as the company gains more experience in the use of those techniques. Day-by-day competence in the employees' task has to be developed.
- If multi-skilling is company policy then employee development is needed to implement training in the new skills.

Both training and development are part of a HRM strategy to which line managers must contribute.

5.4.2 Appraisal and reward

Some form of performance appraisal must be related to training and development, both to identify what kind of training and development is appropriate and also to assess the level of achievement. This may be embraced enthusiastically by achievers but is a difficult and stressful activity for both superior and subordinate where performance is sub-standard.

For the achiever appraisal may be welcome to confirm that they are doing a good job. Not all achievers may be judged to be doing a good job because those achievements may not be to the superior's liking, even if they are in the best interests of the company. Entrepreneurial activity needs to be stimulated in the organisation. New ideas bring about the developments which lead to competitive advantage but they can make life very uncomfortable.

For the low achiever the appraisal may appear to be a threat. Even though the outcome may be suggestions as to how training and development could help, there is the uncertainty of how the next appraisal will rate performance.

A further problem is that achievement is often by teams rather than individuals. Is the team to be held accountable and individual performance in the team to be left for the team to assess? Or is the manager to whom the team is responsible to judge individuals? There has to be accountability of the individual as well as the team but with 'empowerment' does the team have to have a leader who will make this judgement?

Subjective appraisal is perhaps easier to make than objective appraisal. The appearance can be attractive and conform to the expected norms although the outcome may be indifferent. On the other hand the outcome may be excellent but obscured by a behaviour pattern which is displeasing to the appraiser. Agreeing objectives between superior and subordinate, or between a superior and a team, is difficult. Operational objectives are more readily defined than strategic objectives but are of a lower order of importance. Requirements change over time and maintaining up-to-date targets is a challenge. These were problems in MBO.

Evidently appraisal has to be formally reviewed at some time interval, frequently annually. This should not be the occasion of the assessment, nor should it be the first that the subordinate knows of the rating that will be given. Appraisal has to be an ongoing activity, formally recorded at intervals with recommendations for training, development, and in the author's view, for reward and promotion. It should be based on agreed criteria that can be satisfactorily assessed by both superior and subordinate.

The problem of linking reward with appraisal has to be addressed. If performance in the period under review has been above average then there will be the expectation of reward in the current job or promotion. The level at which jobs are rewarded allows, or prevents, a company to attract staff in competition with other companies. For each job, or, more likely group of jobs, there needs to be a salary band which permits reward to be matched to performance. The basic level of reward and the width of the band are

The Management of Human Resources

important if the company is going to retain staff in whom considerable sums have been invested in training and development.

Greenfield sites provide an opportunity for recruitment of personnel to a specification which is matched to the needs of the manufacturing operation. Appraisal gives the opportunity to remedy problems which arise from wrong selection at an early stage.

Brownfield sites present historical problems. These can arise because appraisal has not been carried out and there are job-holders who have been allowed to develop sub-standard performance over a period of time. It is very difficult to rectify this situation. There are also personnel who were adequate operating in a previous generation technological environment who are not able to fit in to the changed environment.

Reward can take the form of promotion, pay, benefits and job satisfaction. Gallie and White (1993) report that 'about 40 per cent of those having appraisals said the "reports and appraisals" influenced how hard they worked' while '70 per cent said that "targets" influenced how hard they worked'. Perhaps the 'Hawthorne' effect is still very strong. Appraisal and targets can set a discipline that is accepted if there is a strong company culture. There is no doubt that norms of performance vary between companies.

There are significant rewards from job satisfaction. Job structuring gives more satisfying work than the excessively specialised tasks under scientific management. There is satisfaction from being an effective member of a team and there can be satisfaction in working for an effective organisation. Ultimately, however, there is the expectation that material reward will be made available for performance, especially if that performance is contributing to the profitability of the company.

Examining these tangible rewards it becomes apparent that promotion will be more difficult to achieve in the new style manufacturing organisation with teams responsible for their own tasks and a flattened organisation where the number of managers and supervisors has been reduced. This is not to deny that promotion can be a reward for exceptional performance but it is not available to the degree that previously obtained.

Merit payment related to appraisal can be seen to be a satisfactory reward. Individual bonuses, where the individual can determine the level of production work well. Group bonuses where the group can influence the level of output, and where the group is balanced, can also work well. Fringe benefits and profit-related pay have much less impact.

With training and development emphasising improved performance in the current job rather than preparing personnel for promotion, there must be thought given to a superior grade of operative based on ability and competence.

Mueller (1992) reported that this takes place in the automotive industry. In Britain Ford has created 'Integrated Manufacturing Specialists' who receive on- and off-the-job training in a variety of skills such as maintenance, CNC appreciation/programming, statistical knowledge and team working. The reward is a higher pay grade.

Developing the human resource

Case Study Extracts 5.4 and 5.5 examine the approach to reward and flexibility taken by two of the case study companies. Each approach is matched to the needs of the company. In neither case is it thought that there is a final resolution of either the reward system or work flexibility. Continuous improvement is the aim.

5.4.3 Flexibility

Flexibility in the workforce allows lower manning levels with operatives able to look after different machines. Flexibility also provides for variation due to model mix so that workers not needed on a line where demand is low can be transferred to a line where demand is high. Volume variation is much more difficult to manage. There have been the suggestions of core employees to be supplemented by temporary labour to cope with peaks. This is a solution which has been used in, e.g. the knitwear industry where there is a pool of trained female workers who are no longer in full-time employment but available for part-time work. Another method for seasonal variation is the annual contract where permanent staff work a number of hours in the year distributed at the employer's discretion.

5.4.4 Leadership

To get the trained, flexible, committed workforce needs a very high level of leadership. This is provided in Japanese companies by a management

Case Study Extract 5.4 Flexibility and reward at Stanley Tools (Case 6)

> Stanley Tools has retained PBR where work is repetitive, relatively simple and output is operative controlled. Job evaluation has reduced the number of grades to four and there is flexibility between the grades subject to the operative having been trained to the level. Thus operatives can be paid a higher rate if they are transferred. Motivation to be transferred from areas which are short of work to areas which are short of operatives is provided by the structure of the PBR system, There is no average earnings payment for non-productive work but a reversion to base rate. The only way an operative can earn bonus is to transfer to an area where productive work is available.
>
> Where output is dependent on teamwork in a cell group bonus is paid on output. Where output is machine controlled bonus is based on keeping the machines operating.
>
> While operative flexibility is promoted by multi-skilling operatives, management has reservations as to how far multi-skilling can be taken for maintenance workers as machines become more sophisticated in respect of electronic control.

The Management of Human Resources

Case Study Extract 5.5 Flexibility and reward at TSCo (Case 7)

TSCo started the auto-line with operatives on hourly payment bases on the average of the batch line bonus. The team was multi-skilled and took on responsibility for simple maintenance. Each shift on the auto-line operated as a team without a team leader. As output increased on the auto-line it was evident that output per operative was higher than on the batch line. Workers' efforts in improving the output of the auto-line led to a demand for increased pay.

This led to a reconsideration of factory pay and reward. A scheme has been introduced based on value-added and paid monthly. The bonus is paid on springs produced per paid hour. This encourages workers to achieve production in normal time to maximise bonus. The auto-line bonus scheme is shown in the figure below.

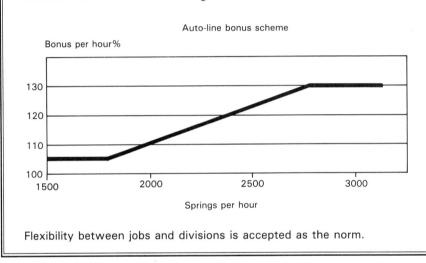

Flexibility between jobs and divisions is accepted as the norm.

determined to solve and eliminate problems. The workforce feels that it has the support of the management. Management pursues the goals of waste elimination with a single-mindedness which builds up a cohesive company culture. The culture is that of an integrated company focused on the customer. British management does not have to imitate Japanese management but it has to have a clear idea of where it is going and set out to reach that goal in spite of difficulties that may arise.

The concept of HRM being strategic is the important message. If the current vogue for TQM is really to be the strategic drive in HRM then management has to identify how such a strategy is to be formulated and, perhaps even more importantly, how it is to be implemented.

■ 5.5 Conclusion

HRM needs a positive strategic leadership from a management that is prepared to regard itself as a part of a unified team concentrating on developing a competitive company.

The relationship with trade unions can and must be developed by the management in a positive way. This is helped by the current climate of improved management/union relationship that has led to a decline in the confrontational stance which obtained previously.

This book is being written at a time when Britain is emerging from a recession with a high level of unemployment with the pendulum of power swung towards the management. It would be unwise to build a strategy on the basis that management can take a compliant workforce for granted. HRM strategy must include 'respect' for the worker if it is to achieve long-term success.

If companies are to have flexible, skilled and well-motivated workforces there needs to be a level of training and development higher than has been achieved in the past. Government initiatives help but they are not the answer to providing such a workforce. Manufacturing companies have to take the initiative and look on government aid as supportive of their efforts.

Success of the manufacturing company depends upon competitive products and processes but these will only be developed and maintained by company personnel who are well trained, committed and motivated. The phrase company personnel has been used to suggest an integrated workforce. Some of the workforce will be managers having responsibility for the work of others. All will have a responsibility for the work of the company which is to satisfy the customer and, as a result, satisfy the company's needs for profit to:

- Reward the workforce.
- Reward the investors.
- Provide capital for the renewal and development of resources.

6 Finance and Control

This chapter will consider:

- The provision of finance for the development of the manufacturing company.
- The control of the business:
  - cost control measures
  - non-financial quantitative and qualitative data.
- Decision making.

■ 6.1 The provision of finance

The sources of finance are shown in Figure 6.1.

A major criticism is made by Hutton (1991) who compares financing in the UK with Germany. Hutton considers that UK companies, compared with German companies, have to pay between two to three times as much from their profits in dividends. As a result investment tended to be made in low-risk indigenous businesses protected from international competition. Figure 6.2 shows investment growth plotted against tradeability. Tradeability is the extent to which the sector competes directly with exports or imports.

The analysis suggests that the biggest growth in UK investment is taking place in service and leisure industries such as banking, communications, hotels and distribution. This analysis gives reasons for 1 per cent growth in UK manufacturing between 1970 and 1990 compared with between 29 and 95 per cent for the balance of the G7 countries. In all the G7 countries the highest growth in GDP is outside manufacturing. In the UK it is only outside manufacturing that there is growth at all.

Within manufacturing industry there is a special problem of low investment and short termism compared with other industrialised countries. The

The provision of finance

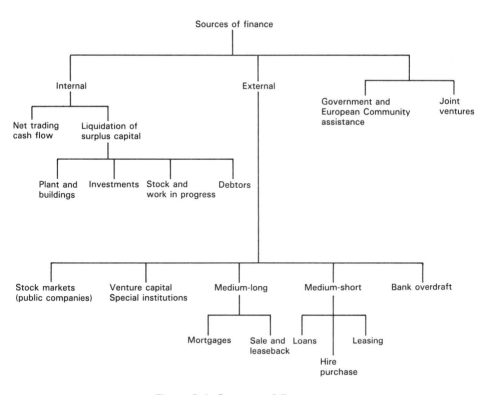

Figure 6.1 Sources of finance

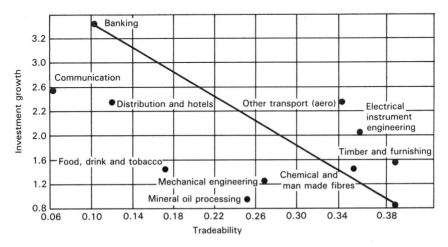

Figure 6.2 Tradeability and investment growth in UK business (1979–87) (*Source*: Muellbauer and Murphy in Hutton (1991))

Finance and Control

level of investment in manufacturing industry during the period 1980 to 1990 for four other industrialised countries is shown in Figure 2.7. It can be seen that the investment exceeds the UK investment by between 44 per cent and 171 per cent. The problem of short termism was discussed in Chapter 3 where it was identified that operational pressures for budget achievement concentrated management effort on short-term profit. These short-term operational pressures were reinforced, in the case of public companies, by the need to maintain the share price which could only be achieved by consistently good annual results.

Allen Sykes in an article in *The Times*, 22 July 1993, commented on a leaked report from the Department of Trade and Industry which showed that British manufacturing was in a poor state with little chance of being competitive for decades. The list of causes included short termism attributed to institutional investors whose goal was to get a high level of return for the institution. If a company did not perform well the shares were sold and the proceeds reinvested. In Sykes' view effective corporate governance requires 'committed and knowledgeable long-term shareholders, managements with the preconditions and incentives for long-term performance, and with such managements properly accountable to the shareholders'.

If DCF is used, which is the theorists' favoured approach for judging capital investment, then Bhimani and Bromwich (1992) suggest that DCF techniques bias against distant returns which are more significant for AMT and against plant- or company-wide benefits. Surveys suggest that payback is, in practice, the commonest justification for capital investment with an expected payback generally being set at two years, which is very demanding. It is doubtful if such a criterion fits into any business strategy other than survival.

6.1.1 Investment and information systems

Several problems arise in evaluating investment in manufacturing.

The interdependence of the manufacturing process restricts the value of evaluating isolated new plant investments. The costs and benefits directly associated with a single investment may not evaluate the true worth of such an investment until taken in conjunction with associated investments.

The difficulty of evaluating intangible benefits arising from investments was outlined in Chapter 2, Section 2.3.4. Strategic approaches to investment decision are more likely to evaluate intangibles than investment decisions taken on an operational basis. Plant replacement decisions taken on a brownfield site are likely to be operational decisions. Greenfield decisions of necessity have to take a view of intangibles. However, when brownfield investment decisions involve new product introduction then intangibles are more likely to be taken into account. Case Study Extracts 6.1, 6.2 and 6.3 shows examples of a brownfield site operational decision, a brownfield site new product introduction and a greenfield site implementation.

The provision of finance

- Case Study Extract 6.1 shows an operational decision involving tangible benefits with a payback of one year.
- Case Study Extract 6.2 presents a strategic decision based on both tangible and intangible factors. There was both return on investment and market share goals.

Case Study Extract 6.1 Investment at Rolls-Royce – a brownfield site (Case 5)

1. Objectives of the advanced integrated manufacturing system (AIMS):

   - To cut work in progress by two-thirds.
   - To compress production leadtimes from 26 weeks to 6.
   - To increase manpower savings by over 40 per cent.

2. Criteria of success:

   - To achieve sufficient savings to cover the project investment cost in its first full year of operation.

   The criteria of success were operational and did not involve 'intangible' factors. This major investment in advanced manufacturing technology (AMT) was made on a stand alone basis.

Case Study Extract 6.2 Investment at Stanley Tools – brownfield site – new product decision (Case 6)

1. Objectives of the new product introduction:

   - To introduce a range of screwdrivers into the domestic and mainland European markets.
   - To satisfy the needs of DIY and Professional users in both wood and engineering aligned trades.
   - To arrest the decline in market share for screwdrivers.
   - To obtain an above average return on investment.

2. Criteria of success.

   - To achieve both market share and return on investment goals.

   The costs and benefits of the investment could be quantified but depended upon the quantification of 'intangibles'. The intangibles arose from the strategic consideration of market shares. To evaluate the risk involved in gaining those market shares, sensitivity analysis was carried out. Management decided that it was prudent to proceed with the level of risk identified.

   The new investment integrated with existing investment for the manufacture of screwdriver bars.

Finance and Control

Case Study Extract 6.3 Hepworth Building Products – greenfield site entry into new sector of the industry (Case 3)

---

1. Objectives:

   - To effect an entry into the concrete drainage pipe industry.
   - To obtain 20 per cent of that sector within five years.
   - To satisfy the following:
     Capital cost   – £14.7 million.
     Sales revenue per annum – £21.9 million (after five years).
     Profit per annum   – very good.
     DCF after tax   – in excess of 20 per cent.
     Payback   – five years.

2. Criterion of success:

   - Viable entry into a new sector of the drainage industry in which Hepworth was a leader.

The success of the investment involved evaluating a range of 'intangible' factors:

- Anticipated growth of the market sector.
- The competitive advantage of the distribution system and the sales force.
- The assured product quality and the incorporation of a fixed jointing ring.

These factors were evaluated on a strategic basis and the project was long term involving a five-year payback period.

The project was a 'business' project and the new factory was to form an independent unit. However, marketing and distribution of the concrete products would capitalise on Hepworth's existing strengths.

- Case Study Extract 6.3 demonstrates a strategic decision to effect an entry into a new market. Intangible factors assumed considerable importance and the investment was regarded as longer term with a five year payback.

The strategic element of the decision increases from operational brownfield site to an entry into a new market via a greenfield operation. The intangible benefits weigh more heavily as the appraisal becomes more strategic. While difficulties remain in quantifying the intangibles, 'what-if' approaches and sensitivity analysis enable management to make a judgement on the risks involved and the prize to be won. Primrose (1988) takes the view 'that every benefit which can be identified can be re-defined, quantified and included in an investment appraisal'.

The decisions in the case study extracts progress from an operational, stand alone decision to an integrated business decision. Problems of how to achieve integration can arise in companies where a series of operational decisions has been taken.

Investment decisions suffer from problems of how realistically costing systems can give the cost of producing a product and the estimated savings. This becomes increasingly difficult as technology increases fixed cost and reduces variable cost, such as direct labour. The traditional systems of job costing and standard costing are, in many cases, not adequate in this new situation. As a result academic and other theorists put forward new systems to deal with the problems of identifying true product cost. This may be an insoluble problem on which a lot of time and effort could be expended with little return. Claret (1987) considers that 'there is no such thing as a true total cost for a product'. There has to be sympathy with this view since some costs, of necessity, have to be allocated.

Major problems exist for manufacturing industry in applying the proposed new systems:

- Systems are needed for making decisions as well as for control. The Japanese have adapted systems which appear largely to satisfy these needs as will be discussed in Section 6.2.1. The adaptation appears to be more the way in which management uses the systems rather than in changes in the system, for example overhead recovery based on labour hours appears to be the method favoured by Japanese companies.
- There is a necessity to produce annual financial accounts. If the proposed system diverges from a system integrated with the production of financial accounts in the accepted format, then there will be the necessity of operating two systems. Reconciliation between the two systems, previously managed satisfactorily, will become more difficult. In many leading European companies this is not considered to be critical, particularly as stock and work in progress are reducing significantly in many companies as management attitudes to manufacturing change and systems such as JIT and TQM are introduced.
- Existing computer software systems are written for conventional accounting and cost control systems.
- There is no agreement on a single new system. The variety of systems will be reviewed below. If a company decides on a new system which it considers is suited to its operations, considerable time and expense will be involved in the change. Only after experience has been gained will the company be able to evaluate whether it has gained benefits from the change. Reviews of the practices of the most successful manufacturers, the best of Japanese companies, do not show major departures from traditional systems. Rather, they show different approaches to the use of traditional systems.

Finance and Control

6.2 Control of costs in the business

6.2.1 The Japanese approach to cost control in consumer durables

According to Hiromoto (1988) accounting systems are used to 'motivate employees to act in accordance with long-term manufacturing strategies'. He considers that accounting supports market-driven management. When a new product is to be put on the market a target selling price is established based on what the market will accept. A target profit margin is specified which reflects the strategic plans and financial projections of the company. The difference is the allowable cost, which becomes the target cost.

This will involve techniques such as value engineering and interaction between design, purchasing, parts suppliers, and the shop floor. Thus there are two costing activities:

- Cost reduction.
- Cost control.

Cost reduction

Cost reduction, in many cases, is concentrated at the design stage because 80 per cent or more of costs are built in here and production parameters are set at that stage. Nevertheless as Hiromoto (1988) instances in the case of a new car: 'This cost is a starting point, however, not an ultimate goal; over the course of the year, it is tightened monthly by a cost-reduction rate based on short-term profit objectives. In subsequent years, the actual cost of the previous period becomes the starting point for further tightening, thereby creating a cost reduction dynamic for as long as the model remains in production.'

Cost control

Cost control is the control of production activities to the standards which have been achieved by cost reduction. In an interview with accountants at Daihatsu Kogyo, Monden (1989b) was told that the direct cost of products for each shop in the plant was broken down as a cost target figure. 'Our budget period is for one year, but only the first six months are really set up with precision. The second half of the year is comparatively loose. We use the rolling method to correct the six-month periods'.

Rhefeld (1990) became familiar with this system at Toshiba and introduced it into the seven divisions of Seiko in America. The review of the budget is not operational but strategic testing the underlying premises of the company's or SBU's activities This is the same approach adopted by, e.g. Hitachi Consumer Products UK.

Control of costs in the business

The use of management control to gain long-term advantage

A conclusion which emerges from examining the way in which Japanese companies approach management control is that companies are looking for systems that motivate employees to pursue long-term competitiveness. The attitude of continual improvement instilled in workforces of Japanese companies is in line with the strategic use of budgets to encourage personnel to look beyond narrow functional boundaries. It appears that Japanese manufacturers are intensely concerned with success in the market place and are determined to achieve this in spite of defects in accounting and control systems.

Bromwich and Bhimani (1989) comment that the efficacy of Japanese accounting techniques 'may rely on a process of informal interaction and communication which is not always understood in the West'. Costing systems are used to influence behaviour. For example, target costing is a goal to be achieved in the knowledge that tighter targets will then be set in line with a philosophy of continuous improvement.

The twice yearly budget gives opportunities for a strategic review as managers are forced to consider the changed circumstances which will require a revised budget.

Using costing systems as influencing systems rather than as 'true cost' systems can alter the appreciation of the need for change. The influencing role of management accounting thus dominates its informing function. This explains why, in capital intensive environments, Japanese companies continue to use labour hours rather than machine hours as the basis for overhead recovery. The managers in this situation will be encouraged to reduce direct labour and promote automation. More reliance is placed on non-financial measures as the Japanese strive to achieve ever better quality and value for money in the market.

6.2.2 The Western approach

The Western approach is quite different from the 'influencing approach' highlighted in the previous section. The defects in management accounting and control systems have been highlighted by academic accountants, debated by accounting bodies and are currently being examined by accountants in industry. How much such measures will stimulate managers is uncertain. How they will be translated into terms which will stimulate all employees to produce constant improvement can only be conjectured.

The problems stem from two different although related shortcomings:

1. The first arises in the multi-product company where, it is argued, traditional accounting systems wrongly allocate costs between products. If costs are allocated on a labour hour basis then small volume products and similar large volume products, which, per unit, have the same direct labour hour content will have the same overhead

allocation. It is claimed that in reality the small volume products have a much higher overhead per unit based on a disproportionate need for changeover/set-up, purchasing, administration, etc.
2. The other shortcoming is brought about by the decrease in direct costs and the increase in overheads. This results in the suggestion that overhead recovery on the basis of labour hours is inappropriate in a largely machine-controlled environment although, according to Hiromoto, this does not appear to be a problem for companies such as Hitachi. While the Hitachi VCR plant is highly automated further reduction in direct labour is needed to promote long-term competitiveness. 'Allocating overhead based on direct labour creates the desired strong pro-automation incentives throughout the organization'. This again focuses on the Japanese use of costing systems in an influencing role.

A final criticism is that management accounting systems do not produce information that will enable management to control overheads.

The impetus for new accounting systems came from academics in the US and has subsequently received wide exposure in the UK. A question that must be raised is whether the same problem is present in all manufacturing industry and which of the suggested techniques should be adopted.

CAM-I (1988b) clearly identify the manufacturing group they are examining by the title of their book *Cost Management for Today's Advanced Manufacturing*. The cost management system discussed in the book includes activity costing but includes elements from other systems outlined below. The perceived advantages are:

- Continual improvement in eliminating non-value-added costs.
- Activity accounting.
- Externally driven targets, including target cost.
- Improved traceability of costs to management reporting objectives.

Surveys that have been carried out in manufacturing companies show that activity based costing (ABC) is the most popular of the new systems. Nevertheless the adoption of ABC is quite low. Miller *et al.* (1992) in their global benchmarking survey find that in the US, UK and Japan the past pay-off from ABC is rated 25 out of 26 activities. In the future, expected pay-off from ABC is ranked 21 in the US, 19 in the UK and still only 25 in Japan.

This reflects difficulties in the West in implementing new systems compounded of problems of conviction and of changing from existing systems. Japanese manufacturing companies, as stated, prefer to remain with traditional systems but change the emphasis in the way they are used.

6.3 Description of the techniques

6.3.1 Activity based costing (ABC)

ABC is concerned with how costs are allocated to individual products in a multi-product environment. ABC attempts to identify the costs arising from greater product diversity. Costs are allocated to products in proportion to the activities and resources consumed. Activities will be consumed in producing, marketing, selling, delivering and servicing products. The amount of activity will vary between products.

ABC focuses on what factors affect the cost of the product. Cost drivers are factors which determine the work load and effort required to perform an activity. Thus the primary cost drivers for manufacturing overhead are not the physical volume of production, but the transactions that require the exchange of material and information.

For instance the overhead associated with purchasing should not be allocated to products on the basis of the volume of product made but on the basis of the demand created on the purchasing department by that product. In order to establish this it is necessary to identify the cost drivers of the product that generate activity in the purchasing function.

Bellis-Jones (1992) states that in the automotive industry only 37 per cent of costs are controllable and of this 27 per cent lie in overheads and 10 per cent in direct labour, raw materials and bought-in components. The main accounting effort goes in to producing information about direct costs. To produce usable information for managers to control overhead costs there is still a problem of relating overhead costs to products. This is the area where ABC costing is said to be accurate. The effect of applying ABC costing is shown by Bhimani and Pigott in Table 6.1.

The application of ABC was at Evans Medical and, as can be seen, the contribution made by each of the four lines altered considerably. While management was unable to alter the pricing on the existing lines the revised information lead to the abandonment of one line.

Table 6.1 ABC costing at Evans Medical (standard v ABC product cost comparison)

	Standard cost	ABC cost	% change	Pack price
Erythrocodil tablets (10 mg ×50)	3.13	4.10	+30%	11.10
Amichloride tablets (5 mg ×100)	0.56	0.40	−28%	2.45
Ephendrine tablets (60 mg ×250)	1.67	6.74	+300%	2.66
Codiphine linctus (2 litres)	3.38	4.03	+19%	4.46

Source: Bhimani, A. and Pigott, D. (1992) *Management Accounting Research*

The advocates of ABC consider that wrong decisions are made using conventional costing. In particular, they say that incorrect allocation of overheads results from conventional allocation systems based on some volumetric concept, e.g. direct labour. They argue that high volume repeatable products do not consume more overhead than small volume irregularly made products as conventional allocation would suggest. In fact not only do they consume less overhead per unit but may consume less overhead in total.

However, Morrow and Connolly (1994) warn that ABC has to focus on specific business issues. Additionally absolute accuracy cannot be achieved. Some activities cannot be measured precisely and, for others, the cost outweighs the benefit.

6.3.2 Cost modelling/simulation

With the advent of desktop computers and increasingly sophisticated packages a variety of analysis is carried out. Spreadsheets are a main form of modelling. At a more advanced level, the simulation programme allows more complex models to be constructed. Either will allow for experimentation and a search for the relationships that obtain in practice.

6.3.3 Cost of quality

The overall cost incurred in achieving the desired level of quality in an organisation's goods or services is frequently analysed over:

- Prevention costs. Costs of activities such as quality engineering or supplier assurance which are carried out to investigate, prevent or reduce defects and failures.
- Appraisal costs which are incurred in measuring the level of quality achieved such as inspection, laboratory testing and field testing.
- Failure costs arising because of a failure to achieve the required level of quality such as rework, scrap and warranty.

These costs exist in most organisations but are not reported. When extracted they frequently amount to between 15–20 per cent of turnover.

In practice control is not exercised by financial measures but by other quantitative and qualitative measures. Training and management control are essential to TQM. Cost of quality may be a confirmation of the success but not a main indicator.

6.3.4 Target cost planning

Target costing has already been reviewed in considering approaches to cost control. Target cost planning takes place at the design stage of the product and is a team effort depending upon the commitment of the team, the ability to cross functional boundaries and the autonomy allowed the team for decision making.

The activity is receiving attention in manufacturing companies under concurrent engineering in America and simultaneous engineering in the UK.

The success of target costing depends not only on the autonomy given to the team which has the task of achieving the target cost but also to the determination with which they attempt to achieve the target cost and the product launch on time. A formal 'buy-off' of the design and manufacturing process operates in many leading companies.

6.3.5 Strategic management accounting

Strategic management accounting was defined by CIMA (1982) as: 'The provision and analysis of management accounting data relating to a business strategy: particularly the relative levels and trends in real costs and prices, volumes, market share, cash flow and demands on a firm's total resources'.

It has been argued in this book that the SBU is the level at which strategy is determined and therefore must be the level at which strategy should be monitored. Conventional accounting is inward looking reporting on budget variance, making comparison with previous periods of the same year and the previous year. Simmonds (1992) advocates the presentation of strategic management accounting data in a format shown in Table 6.2. As can be seen

Table 6.2 Strategic accounting indicators

	Volume	Unit revenue	Unit cost
Ourselves			
Current position			
Change in period			
Lead competitor			
Current position			
Change in period			
Compared to us			
Close competitor			
Current position			
Change in period			
Compared to us			
Weak competitor			
Current position			
Change in period			
Compared to us			

Source: Simmonds, K. (1992) Strategic management accounting: what makes it different? *Manufacturing Technology International*

the concept has much in common with benchmarking and the indicators could be extended to include, e.g. turnover per employee and added value. The problems of getting access to data are high particularly for small and medium-sized companies.

In considering strategic management accounting Bromwich (1990) agrees that the focus must be external to the business and should provide information which will allow a company to identify the effects of altering the way in which the business is organised and run. For example, what would be the effects on the business of economies of scale? The increasing use of PC models and 'what-if' options allows management to evaluate possible scenarios on a given set of key assumptions. Modelling of this nature is a powerful way to assist management in decision making and formulating policies.

Hepworth utilised a form of strategic management accounting to enter a new market. The case study emphasises the strategic value of the distribution chain. The CIMA definition emphasises 'total resources' but the lack of integration in much UK manufacture raises the fear that, whether strategic management accounting were to be adopted, or benchmarking, the use of the information would tend to be compartmentalised. This would lead to a consideration of the company as an amalgamation of a series of specialist producers and ignore the benefits that can accrue from economies of scope. For example, Stanley Tools derives benefits over a range of products from applying expertise in moulding, in forging and in packaging.

The critics of SBUs would consider that there is the danger of a company becoming a series of autonomous units incapable of giving synergistic benefits. A matrix organisation is needed to prevent this happening.

The compartmentalisation brought about by functions needs to be broken down so that managers and accountants become managers with a common goal. The strategic direction of the business cannot be guided just by profitability, especially short-term profitability.

Hiromoto (1988) perceives Japanese competitiveness stemming from management accounting systems which 'reinforce a top-to-bottom commitment to process and product innovation'.

6.3.6 Throughput accounting

This technique is related to the concepts contained in OPT that costs are allocated on the basis of throughput times. This can best be summarised in the productivity improvements aimed at by Courtaulds Filament Yarns shown in Figure 6.3.

As can be seen the only product that should be made is that which can be sold. Only money generated by sales is recognised. All future revenue contained in material stocks does not contribute to profit.

Operating expenses similarly are the object of close scrutiny. Thus management is being asked to adopt a different attitude to the way in which

Description of the techniques

```
┌─────────────────────────────────────────────────────────────────┐
│ PRODUCTIVITY                    OPT                              │
│ IMPROVEMENTS                                                     │
│                                 Optimised Production Technology  │
│ Evaluate each decision; will it:                                 │
│                                 Uses three global measures       │
│ • Increase throughput                                            │
│                                 THROUGHPUT – the rate at which   │
│ • Reduce inventory              money is generated by the system │
│                                 through sales                    │
│ • Reduce operating expense?                                      │
│                                 INVENTORY – all of the money a   │
│ SIMULTANEOUSLY                  firm invests in purchasing       │
│                                 things which it intends to sell  │
│    ⇑    ⇓    ⇓                                                  │
│    T    I    O                  OPERATING EXPENSE – all of the   │
│              E                  money spent in order to turn     │
│                                 inventory into throughput        │
└─────────────────────────────────────────────────────────────────┘
```

Figure 6.3 Courtaulds Filament Yarns – productivity improvements and OPT (*Source*: Courtaulds Filament Yarns)

the business is managed. Whether this demands a change in the system of accounts is open to debate. There can be no doubt that there has to be a difference in the values that management use for judging the business.

Waldron and Galloway (1988) suggest measures which will allow management to judge performance, as shown in Figure 6.4.

6.3.7 Backflush relief of inventory

JIT can allow use of backflush accounting. There is a focus on output followed

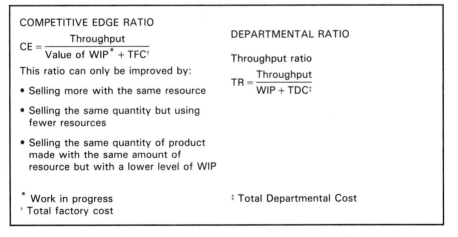

```
┌─────────────────────────────────────────────────────────────────┐
│ COMPETITIVE EDGE RATIO                DEPARTMENTAL RATIO         │
│                                                                  │
│            Throughput                 Throughput ratio           │
│ CE = ─────────────────                                           │
│       Value of WIP* + TFC†                     Throughput        │
│                                        TR = ──────────           │
│ This ratio can only be improved by:          WIP + TDC‡          │
│                                                                  │
│ • Selling more with the same resource                            │
│                                                                  │
│ • Selling the same quantity but using                            │
│   fewer resources                                                │
│                                                                  │
│ • Selling the same quantity of product                           │
│   made with the same amount of                                   │
│   resource but with a lower level of WIP                         │
│                                                                  │
│ * Work in progress                    ‡ Total Departmental Cost  │
│ † Total factory cost                                             │
└─────────────────────────────────────────────────────────────────┘
```

Figure 6.4 Throughput accounting measures (*Source*: Waldron and Galloway (1988))

Finance and Control

by working backwards to inventory without accounting for WIP. Thus cost of goods sold is the standard conversion cost plus the cost of raw materials.

This simplified costing system is only justified if material moves along the production line as sales are made without intermediate storage.

6.3.8 Life cycle costing

Cam-I (1988b) advocates the expansion of traditional cost accounting procedures to include the non-recurring costs that occur during the product development and the after-sales support phases of a product's life cycle. As competitive pressures push suppliers to show an overall operating cost lower than the competition, e.g. airlines, aircraft and engines, then life cycle costing becomes more important.

■ 6.4 The value chain

The value chain is a concept evolved by Porter (1985b) as a way of analysing sources of competitive advantage. Competitive advantage comes from creating value for buyers that exceeds the cost of generating it. Manufacturing companies perform a variety of tasks in transforming raw materials and bought-in products into final products. The value chain considers how a particular company or business unit carries out these activities to maximise the creation of value.

Using the concept in practice presents many of the difficulties experienced in strategic management accounting and in benchmarking. The systems of measurement are not well developed and the data required is not readily available.

Competitive advantage can come from a company's access (or lack of access to resources). For example, Rover and its forerunners, BL and BMC did not have an effective distribution network in the US. Whatever the quality/cost competitiveness of the products they were at a disadvantage compared with competitors. Value can be added by being more effective than competitors in areas of managing the value chain. There are gains to be made in eliminating non-value-adding activities.

The first approach is to identify which activities contribute, or do not contribute, value to the product. These activities may be in the incoming supplies to the company, in the product design, in the methods of manufacture or in the distribution to the customer.

A second area is in the linkages and interrelationships. These may be internal or external. Internal may be between sequential tasks and value added, or between business units. Externally they may be vertical between the company and its suppliers or between the company and its customers. Horizontally they may be coalitions or long-term alliances with other companies.

■ 6.5 Non-financial quantitative and qualitative measures

The discussion of new accounting methods shows that accountants are increasingly aware that the control information they produce has defects when used for control of the business and for making decisions. The solution to the problem requires that the functional divide between management and accountants should be removed. Management realises that flexibility gives strategic advantage, particularly in the durables sector of the market. A major contributor to flexibility is JIT but Mackey (1991) says that Ohno, who helped to develop JIT at Toyota, could only get his system to work when the accountants were banished from the plant. Further, he could not justify JIT using any existing accounting analysis.

There also has to be the recognition that not everything needs to be the subject of financial control. To shut down a line when a defective part is produced and take action to ensure that the defective part will not be produced in the future is an example of non-quantitative control. In the past such an action would not have been taken because the defect would not have been found until inspected at the end of the line.

Many actions such as the reduction of the supplier base, shortening lead-times through the factory or monitoring the satisfaction of promised delivery dates require non-financial quantitative measures.

One way in which Japanese managers retain control of production and identify where improvement is needed is by observation.

Customer satisfaction cannot be gauged just on satisfactory delivery performance but needs contact by senior company managers.

■ 6.6 Decision making

The problems of allocating costs for decision making are the reasons given for management accounting changes, particularly strategic management accounting focusing outside the company on the competitive environment. The intention to build a management accounting system which will focus on strategy is valuable but does not alter the accounting and control systems in use for controlling internal company operations. Undoubtedly there would be a better database for decision taking. Such a monitoring process would keep strategy under review. Nevertheless there is still a need for quantifying the intangibles which will arise from capital expenditure. The intangibles may not be related to what advantage will be gained from the proposed strategy but rather, that if the strategy is not adopted the company will cease to be a competitor.

ABC and throughput accounting contribute to improved decision taking. The former by attempting to unravel the true cost of products and presenting management with a tool for controlling overheads. A better understanding of the allocation of overhead costs can help the decision-making process. Throughput accounting can be particularly beneficial for those companies adopting OPT as the method of resource planning. Its protagonists would claim that it represents the effects of JIT/TQM in the waste reduction which this system aims to achieve.

■ 6.7 Conclusions

Obtaining finance to expand the business appears to be more difficult in the UK than in competitor countries. The emphasis is on the short term which makes manufacturing investment decisions difficult since they are essentially long term.

The Japanese approach to capital investment is long term and strategic. Using target costs and cost reduction as the main ways of building competitive advantage the Japanese have much less concern than the West with the accuracy of accounting systems. The use of the systems for control is, in any case, more an influencing system to drive forward the cost reduction programmes. Budgeting is often a twice yearly activity and once more is used to revise strategy.

The West is concerned with costing accuracy. ABC costing addresses the anomalies which arise in overhead cost allocation in multi-product plants where there is a mixture of large and small volume products being made on the same plant. Strategic accounting is being offered as a way of promoting an outward look for the business. This can also be promoted by strategic benchmarking.

Financial figures are the only means, ultimately, of comparing performance between companies. Nevertheless, assessment of developments which should be made within a company need to take account of many factors which are not part of the financial control system. Similarly cost reduction and cost control depend upon factors other than those which would readily be used by accountants and the finance function. The solution is for the accountants whom Drucker (1990) terms 'bean counters' to count the beans differently so that manufacture is integrated with business strategy. Accountants need to join with managers as a team for managing the company to gain strategic advantage. Drucker sees the necessity to replace traditional accounting systems by a system which uses time as its measurement unit. This still leaves the problem of opinion in assessing the impact on the business of manufacturing changes. Drucker is worried about quantifying opinions.

As overheads become a larger proportion of manufacturing cost the challenge of what basis to use for costing overheads assumes greater importance. How should these overheads be allocated to products? In making decisions about the future how can intangibles be given adequate consideration? Evidence suggests that it is attitude and vision which need changing.

Case Studies

CASE STUDY 1

DSF Refractories Ltd

CS1.1 Introduction

DSF is a company within the BH-F group of companies. The BH-F group is in the business of supplying equipment to the thermal manufacturing industries.

DSF predominantly manufactures high-alumina refractory products. Small amounts of silica and magnesia-based products are also made. Competitors in UK and the rest of Europe are mainly large manufacturers:

- Hepworth Refractories Ltd.
- Dyson Refractories Ltd.
- Didier Werke.
- Martin and Pagenstrecher.
- Radex.
- Steuler.

The process of manufacture involves mineral processing, brick making and chemical casting.

CS1.1.1 Mineral processing

Two operations are carried out in mineral processing:

- The manufacture for sale of crushed and screened raw materials.
- The provision of the raw materials needed by other areas of manufacturing.

CS1.1.2 Brick making

Bricks are refractory shapes.

CS1.1.3 Chemcast

Chemcast manufactures chemically cast products.

CS1.2 Process flow

All these processes start with crushing, grinding and/or screening raw materials based on high alumina. These raw materials are mainly calcined bauxites from Guyana, China and Brazil where refractory grade raw material deposits are found.

Case Study 1: DSF Refractories Ltd

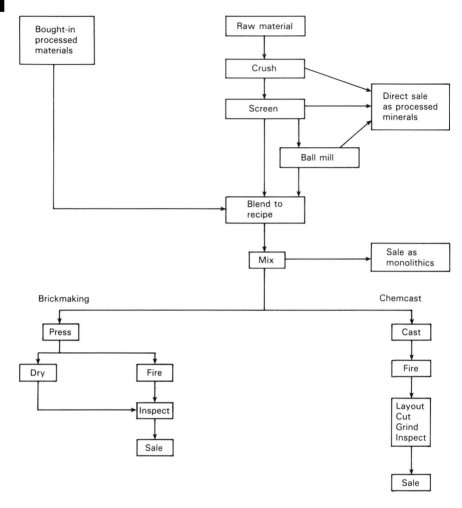

Figure CS1.1 Process route

Other high alumina raw materials used are sillimanite, andalusite and clay materials from various areas of the world.

Bulk tonnages of material arrive in lump form. Other, smaller use materials, arrive already processed and bagged. The process route is shown in Figure CS1.1.

CS1.3 DSF customer base

DSF customer base is in steel, glass and rotary kiln (cement) industries. There is also a general market that includes furnace builders, chemical plant, crematoria and the aluminium industry.

The mineral processing outlets are other refractory producers, welding rod manufacturers, sanitary ware producers and road surface dressing companies.

DSF customer base

DSF is dependent upon the steel industry for over 50 per cent of its turnover. It intends to reduce this dependency by establishing and expanding refractory niche markets in industries such as glass, cement, pulp and paper. Currently DSF is penetrating the rotary kiln market in pursuit of its strategy to diversify away from the steel industry.

CS1.4 DSF business plan

In 1990 DSF prepared a business plan which concentrated on the five goals outlined below.

CS1.4.1 Broader market base

Detailed plans were prepared to identify markets that would allow growth and at the same time reduce dependency on the steel industry.

CS1.4.2 Optimise utilisation of plant

Business would be sought which would suit current production facilities. The object was to delay obsolescence of existing plant. Development of such business had to avoid conflict with long-term growth or product development plans defined in Section CS1.4.1.

A prime example was the surplus capacity for drying chemically bonded products. New applications were being sought in the steel and cement industries and new chemically bonded products would be included in the product development programme.

CS1.4.3 Growth and profitability

A growth and profitability plan was based on the market plans prepared in Section CS1.4.1 which showed:

- Growth in turnover of 80 per cent over the 5 years.
- Approximate doubling of margins.

The importance of markets would alter as shown in Table CS1.1.

Table CS1.1 Analysis of sales by market as a percentage of turnover

Year	1989/90	1990/91	1991/92	1992/93	1993/94	1995/96
Steel	39.6	34.4	30.0	25.8	22.3	18.9
Glass	14.1	18.3	19.2	20.0	20.7	21.4
Cement	10.9	11.8	12.9	14.1	15.1	16.2
Others	12.0	11.2	12.0	12.8	13.4	13.9
Mineral Processing	23.4	24.3	25.9	27.3	28.5	29.6
	100.0	100.0	100.0	100.0	100.0	100.0

Case Study 1: DSF Refractories Ltd

CS1.4.4 Modernisation

Capital expenditure proposals were detailed for additional plant to improve efficiency, improve product quality and increase capacity in line with the requirements of the plan.

CS1.4.5 Reduce dependency on Guyanan bauxite

During the previous 3 years there had been serious supply problems with Guyanan bauxite, a major raw material. Programmes of product development were undertaken based on alternative materials such as Brazilian and Chinese bauxites.

By the beginning of 1993 satisfactory sources of alternative materials were established.

CS1.5 Manufacturing strategy

CS1.5.1 Manufacturing audit

An audit was undertaken of existing manufacturing facilities. Manufacturing was identified as: 'All activities and resources relating to the purchase and conversion of bought-in materials into the required products'.

There was no readily available central record of plant capacities or limitations. To set up a database of plant capacities historical records had to be investigated and discussions held with managers, supervisors and operators.

The aim of the manufacturing strategy was to:

- Attain the order-winning criteria of being a low-cost producer with short leadtimes and a high level of delivery reliability.
- Ensure sufficient and suitable manufacturing resources to satisfy increased forecasted demand.

The manufacturing audit had identified the following problem areas.

Set-up times on presses

A total production time of 11.25 per cent is lost due to die changing. Details are contained in Table CS1.2.

High defective level and cost of quality

Table CS1.3 shows the cost of quality.

Table CS1.2 Die changing time by press

Press	% of production time spent die changing	Hours die changing per year	Typical time per die change in hours
Horn	11.7	352	8.0
Toggle	5.3	287	6.0
Vibro	14.1	2202	2.2
Tamper	9.0	517	2.2

Table CS1.3 Quality cost distribution 1990–1991

Category	Total pounds	Percentage
1 Other grog losses	201,400	29.02
2 Breakdown maintenance	149,000	21.47
3 Crushing & grinding	140,000	20.17
4 Grogged at press	90,720	13.07
5 Credit notes	42,000	6.05
6 Remakes	40,861	5.89
7 Remixes	11,732	1.69
8 Concessions	7,245	1.04
9 Lab re-test	3,600	0.52
10 Refiring	3,300	0.48
11 100% Sort	2,700	0.39
12 Customer complaints	1,560	0.22
Total	694,118	

Significance of failure costs

Annual sales 1990–1991	£11,300,000
Failure costs	£694,118
Total days	240
Average number of employees	182
Failure costs as percentage of sales	6.14
Failure costs per day	£2,892
Failure costs per employee	£3,814

Process time through the range kilns

There are two sets of range kilns. These kilns are interconnected and have to be filled, fired and drawn in sequence. Figure CS1.2 shows a 'perfect' sequence of operations.

- This perfect sequence requires that each kiln is filled to capacity to meet thermal requirements. The result is that excess material is produced that has to be stocked.
- Ambient temperature affects cooling.
- The necessity to fire silica prolongs the cycle time for the kiln containing the silica and also for the preceding and succeeding kilns.

Rejection at the end of the process

Lack of effective in-process quality control leads to the firing of defective material. Thus defective material is only detected at the end of the process affecting both the utilisation of Drayton kilns and production and capacity planning.

High capacity utilisation of Drayton kilns

The Drayton kilns allow for firing at temperatures not available on the range kilns. They also have a much shorter cycle time. Shortages in customer orders due to defective product have to be remade. Such product is routed through the Drayton kilns causing overload.

Case Study 1: DSF Refractories Ltd

Perfect sequence of kiln firing

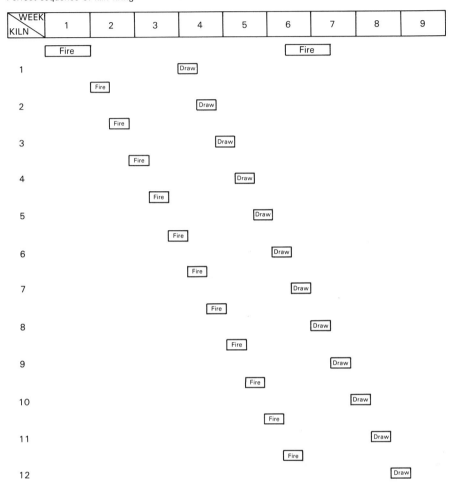

Figure CS1.2 Perfect sequence of kiln firing

Production and capacity planning

Plant is laid out by process. Product generally needs to be manufactured to customer order although there is the possibility of manufacturing some repeat business for stock.

- Capacity planning and control are complicated by process limitations. Press set-up limits change between products.
- Range kiln loading and variation in process time affect production programmes.
- High level of rework interferes with the plans made.

Implementing parts of the manufacturing strategy

CS1.5.2 *Formulation of a manufacturing strategy*

A detailed manufacturing strategy was evolved based on the business plan in conjunction with the manufacturing audit.

Those parts of the plan involving major capital expenditure are not developed in this case study. The recession at the beginning of the 1990s had the two-fold effect of limiting growth in turnover which reduced the need for increased capacity. At the same time, the high level of risk caused by the decline in the economy, together with the limited availability of capital, precluded investment in major capital items.

The areas that could be progressed with limited capital expenditure were:

- Quality management.
- Installation of a new computerised production and capacity planning system.
- Reorganisation of production into three autonomous units.
- Reduction of press set-up time.

CS1.6 Implementing these parts of the manufacturing strategy

CS1.6.1 *Quality management*

DSF achieved registration under BS 5750 Part 2 in September 1988. The need for registration arose from being a supplier to British Steel. The general attitude throughout the works and among senior management was that BS 5750 was a necessary evil that had resulted in increased costs.

In 1992 the emphasis changed to the achievement of total quality. Consultants were brought in to examine the scope of total quality. Figure CS1.3 shows the business improvement programme needed in order to achieve total quality.

DSF business improvement programme

It can be seen that the business improvement programme was to be based on a joint survey of the business process and the quality of service given both inside and outside the company. From these surveys a goal was developed. The goal was DSF as the company personnel would like it to be, see Figure CS1.4

The company had started work on a total quality programme before the consultants were brought in. As a result the concept of TQM was already accepted within the company. Change in attitude in respect of quality had started when control of the brick inspectors and testers had been transferred from quality control to production.

An outcome of the consultants' approach to total quality was that quality improvement was to be introduced by improvement teams who would be responsible for:

- Quality certifications.
- Waste.
- Statistical process control.

These were considered to be important areas to be addressed in improving the business.

Case Study 1: DSF Refractories Ltd

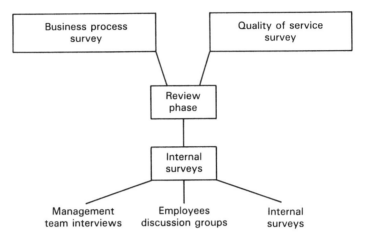

Figure CS1.3 Activities to achieve total quality

CS1.7 Computerised planning and control

Figure CS1.5 shows the business process overview. The existing system only covered sales order processing and stock control. There was a separate accounts system. These two systems were not integrated and thus there were two information databases which were not linked. As a result much management control information was not available.

The additional areas of planning and control that needed to be included to obtain integration and effective control were:

- Production planning.
- Material planning.
- Work-in-progress control.

A decision was made to install a new computer and software that would provide integration between accounting and:

- Sales order processing.
- Stock control.
- Production planning.
- Material planning.
- Work-in-progress control.

The accounting system, sales order processing and stock control modules have been implemented, and the following benefits have already been obtained:

- The sales enquiry system produces information of value to sales managers. As a result they are able to control conversion rates and progress outstanding enquiries.
- Picking lists are generated which improve despatch. Despatches are recorded and automatically update inventory holding.

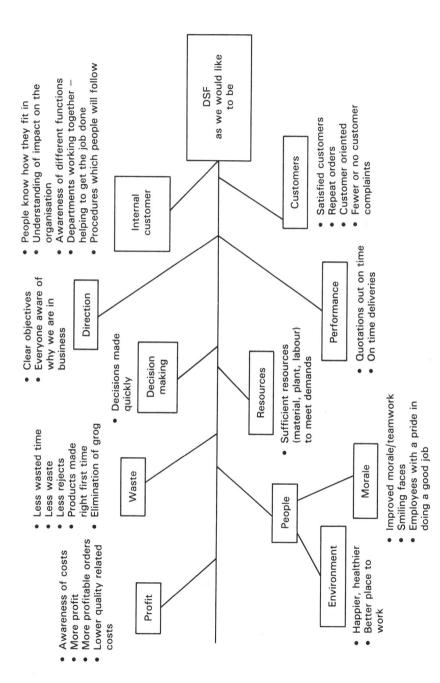

Figure CS1.4 DSF as the company personnel would like it to be

Case Study 1: DSF Refractories Ltd

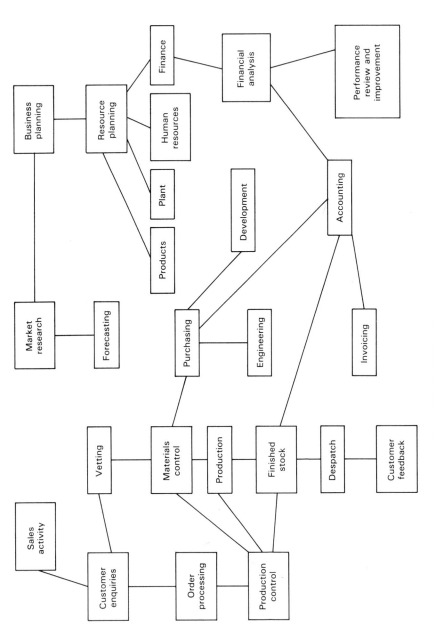

Figure CS1.5 DSF business process overview

CS1.8 Reorganisation of production into three autonomous units

CS1.8.1 Strategic business units

The organisation was restructured into SBUs which would contribute to integrating the business process. SBUs would contribute to the business improvement programme by focusing attention on profit, performance, decision making and customers. It was expected that the change in the organisation would favourably affect the environment, people and morale.

The SBUs were based on the three areas of processing:

1. Mineral processing.
2. Brick making.
3. Chemcast.

CS1.8.2 The goal of the reorganisation

Each business unit would buy raw materials and sell to external customers. Mineral processing would sell raw materials to brick making at agreed transfer prices. Management accounts would be presented in a form that would show the true performance of each business unit. Ultimately each unit would be totally self-contained in respect of plant supervision, maintenance and labour.

CS1.8.3 The extent of reorganisation in 1993

Mineral processing

The involvement of mineral processing in mixing materials for brick making was transferred to brick making. Mineral processing concentrates on preparing raw materials.

The unit operates under the control of a manager who is responsible for both sales and production.

A maintenance fitter was appointed to work full time in the unit. In addition to maintenance the fitter became a production worker which gave him an insight into production problems. Production workers were being trained to do their own maintenance. Screen wear was controlled by the unit and when the limit of wear was reached production workers (not maintenance) carried out the change.

A problem for this unit had been the processing of Chinese YQ bauxite. While the purchase price of this material was lower, investigations proved that the total cost of preparing finished product was higher than for alternative materials with a higher purchase price. As a result YQ was being phased out.

The change from the Chinese YQ bauxite had a beneficial effect on plant uptime. The reorganisation of the unit also increased uptime. As a result there was an interest in and determination to achieve even higher standards of uptime. The drive to reach higher standards was supported by a revised pay system. The payment by results system had been changed to a flat rate system but the rate was higher when the plant was producing than when it was down.

Case Study 1: DSF Refractories Ltd

The outlets for product are now:

- Brick making which now carries out its own preparation and mixing.
- Outside sales.

Brick making

This was responsible to the works manager. The unit was divided under two superintendents:

1. Preparation and mixing together with brick making.
2. Firing.

The unit was more complex than mineral preparation. The changes taking place in the production planning and control system would help scheduling. Scheduling would also be helped by reduction in press set-up time which is outlined in Section CS1.9.

There had been a reduction in brick damage by attention to pallet stacking. This shows an annual saving of £30,000.

Chemcast

This was under the control of the works manager. Although a self-contained production unit by the nature of its process, firing of the product had to make use of the kilns controlled by brick making. Provision of dedicated kilns was planned but capital expenditure would not take place until the economy revived.

The main change to this unit had been the transfer of pattern making into the production area. A problem had been identified that moulds were made for ease of pattern making not ease of chemcast manufacture. The transfer of pattern making to the production area was expected to make the pattern makers more aware of production problems. The intention was to reinforce this awareness by interchanging some grades of pattern making and production workers.

CS1.9 Reduction of press set-up times

Reduction of press set-up times was a part of the DSF business improvement programme. The manufacturing strategy had shown this to be an area where production time was lost. Press set-up times are shown in Table CS1.2.

A working party was established to investigate ways of reducing set-up times. The working party established three areas of investigation:

1. Organisation of distribution, i.e. how to make available all tooling at the press when it was required.
2. The method of die changing using existing tools, equipment and manpower.
3. Re-engineering items or systems or both to effect improvements.

Single minute exchange of dies became a target that was considered to be achievable. Having all tooling at the press when required was a realistic target.

Reduction of press set-up times

Production programmes were sufficiently stable for the day's programme of die changes to be fixed in advance.

The initial thought was for an indirect worker to service die changes. However, considerable gains were made by installing gantries with electric hoists in the die shop. As a result press operators bringing dies for exchange could unload and load dies rapidly. Die shop operatives were able to lay out a day's die changes in a kitting area. The handling system increased the productivity of die shop operatives by 30 per cent, reducing overtime and improving the availability of refurbished dies.

A further achievement of the working party was to standardise the machine beds so that all dies would fit all machines without additional setting operations.

Future progress will be to change dies as a cartridge. This will ease the problems of die location. Faster methods of securing dies in the machine were being investigated.

The reduction of time in having the die available at the press and shortening the changeover time by better methods were expected to get die change to single minutes.

CASE STUDY 2
Fisher Controls Ltd

CS2.1 Product

Fisher Controls designs and manufactures process control systems. These systems are applied to process industries such as:

- Pharmaceuticals.
- Pulp and paper.
- Food and beverages.
- Metal and mining.
- Hydrocarbon processing.
- Oil and gas production.
- Industrial effluent treatment.
- Power, etc.

The systems are built from subsystems or units which contain a configuration of printed circuit boards designed to satisfy customer requirements.

CS2.2 History

At the time this case study was written Fisher Controls at Leicester was a part of Fisher Controls International, Inc. The design of the control systems originates in America. In 1988 printed circuit boards (PCB) manufacture was located at Leicester. The decision to consolidate PCB manufacture was made on the basis of an over-capacity in production facilities between the US and the UK.

Leicester was selected because the operation had moved to a new site, had a lower labour cost than the US, was much less unionised and was seen as an achieving plant located in a technologically developing area. Significant tax credits had accrued at Leicester.

CS2.3 Manufacture

Manufacture is a two stage process. The first stage is the design and manufacture of PCBs. These are subsequently incorporated into cabinets and consoles to form systems built on a jobbing basis to customer specification.

The case study will concentrate on PCB manufacture. In 1992 there were some 230 different PCBs in current production and the annual quantity manufactured was 80–100,000 boards.

The sequence of PCB manufacture is to receive bare boards, axial com-

Manufacture

ponents, integrated circuits and other components into stores. The incoming materials are subjected to a variety of tests before being released to production.

Over time the range of components tested at Fisher Controls has reduced and will continue to reduce as vendors show that they can achieve the standards that Fisher Controls require.

The commencement of improvement in vendor quality started when Fisher Controls rejected batches of components which did not meet the procedures detailed below. Working closely with vendors has resulted in vendors improving process control and testing so that approval can be given for supplies from these vendors to be accepted without further testing.

CS2.3.1 *Components*

These are stored in a carousel which is controlled by a slave computer linked to the MRPII system. Components are withdrawn from the carousel based on bills of materials for a particular PCB. The program is fed through the computer and the appropriate components withdrawn from the kitting area.

- Axial components which are resistors, capacitors, diodes, etc. are fed through a sequencer which is controlled by a computer and driven by the program for the particular PCB. The sequencer tests for the correctness of the part in respect of tolerance and functionality and packs the part into a bandolier in sequence. In case of failure the operative has a mini-store of components from which replacements may be obtained.
- IC components are sample tested for reliability on receipt. One hundred and twenty-seven are selected from each batch up to a quantity of 3000. If one fails, more tests are carried out, if two fail, the batch is rejected. A bill of materials is printed for the PCB being kitted and tubes of components are selected from the carousel and placed on the kitting trolley which has numbered locations corresponding to the insertion machine location.

CS2.3.2 *Process*

The first stage of PCB manufacture is machine insertion of components. The dual inline packages machine inserts the integrated circuits (ICs) followed by the variable centre distancing machine which inserts axial components. The machines can take variable sizes of circuit board. Fixtures are mounted on the machine, the programme for the PCB is fed into the machine and the machine enters the components automatically.

The next stage is hand mounting of components such as connectors, transformers, switches and capacitors.

From this stage the PCBs go to an automatic flow soldering machine. It may be necessary to mask some areas of the board for second stage insertion and hand soldering.

When completed, the boards go to test. 'In circuit' testing monitors components and links between components. Functional testing identifies the board's capability of performing to its design function. Environmental screen testing (run in) takes place at temperature, e.g. $50°C$ for twenty-four hours. If the board

Case Study 2: Fisher Controls Ltd

passes this test there is a high statistical probability that no failure will occur during the next 5–7 years.

Boards are despatched to Unit Assembly or to the customer. There are four major European locations which receive PCBs assembled in unit cabinets to customer specification. The American location receives PCBs to assemble into unit cabinets.

CS2.3.3 Problems of manufacture

Six months after the transfer of manufacturing to the UK there was a major problem of supplying product to the worldwide operation.

In 1989, Chuck Kraemer was appointed manufacturing director to sort out the problem. He was from Austin, another Fisher plant in America. His approach was to identify a manufacturing strategy. While this was in the process of development he took steps to make the organisation structure flatter.

The manufacturing strategy was prepared against a company background of vision statements and the worldwide manufacturing long-range plan.

CS2.4 Strategy

CS2.4.1 Company strategy

The Fisher worldwide vision is:

- We will view the world through the eyes of our customers.
- We will be a global, market driven, quality enterprise.
- We will create a competitive advantage for our customers through applying products and services.
- We will create an environment in which employees and other stakeholders can rely on the integrity of our commitments and where all employees are given the opportunity to realise their full potential.

CS2.4.2 The long-range plan, 1990–99

The long-range plan was based on business environment assumptions and product direction assumptions. An increase in business was forecast over the period of the plan. Manufacturing long-range business objectives were:

- *Internal*
    Increased return on capital (asset management).
    Inventory reduction.
    Supplier management.
    Decreased cost (cost management).
    Manufacturing productivity.
    Supplier management.
- *External*
    Achieve quality leadership.
    Exceed customer expectations for delivery, quality and reliability.

CS2.4.3 What Fisher will manufacture in the mid-1990s

Fisher will make those things where it provides cost effective solutions.

- Through hole boards. Where this is cost effective and strategically important.
- Surface mount technology (SMT).
- Electro-mechanical assembly. Where this is cost effective and strategically important.

CS2.4.4 Manufacturing strategy

The background to strategy development:

- The manufacturing operation had to improve if it was to survive. Competitors' performance was known to be much higher than Fisher's. The competitive stance of Fisher depended upon being able to support system development which in turn depended on the flexibility of PCB manufacture. As a result it was decided to develop a manufacturing strategy which would improve the manufacturing operation.
- Chuck Kraemer was a charismatic leader who adopted an open style of management. His organisational changes had demonstrated his intent to resolve the problem.
- There was no resistance from the unions since there were no major issues to contest. The move to the new site in 1986 at a cost of £6m showed commitment in the workforce's eyes. Common status in hours worked, facilities used, monthly pay, etc. gave the workforce the feeling of being an integral part of the company.

The evolution of manufacturing strategy

The manufacturing strategy that evolved addressed five areas:

1. People involvement.
2. Total quality.
3. Supplier partnering.
4. MRP/JIT.
5. Loss prevention.

The question under each of these headings was: 'Where should we be in 2–5 years?' Targets were set based on what was needed in comparison with competitors' achievements and on what the company identified as necessary, e.g. an 80 per cent reduction in inventory holding.

At this stage management did not know if the targets could be achieved but they accepted that these were the targets which had to be met if Fisher was to remain competitive.

Strategy implementation

There was extensive consultation with the workforce although the evolution of the strategy was firmly in the hands of the management. The company was at a

Case Study 2: Fisher Controls Ltd

learning stage in identifying a manufacturing strategy. A group of some 60 personnel, management to shop-floor operatives met off-plant to identify problems and propose solutions.

Management was also unable to foresee some of the outcomes. The group of 60 raised the possibility that the strategy being developed would result in redundancies and were assured that 'the changes would not lose jobs'. Nevertheless, at the beginning of 1992 redundancies took place.

The strategy and how it was to be implemented were communicated to manufacturing personnel via the management/supervisory structure. Although consultation had taken place, the management structure was considered to be the appropriate mechanism for putting the strategy into operation.

The group of 60 spent in total at least one week in off-plant meetings. The whole of manufacturing spent at least one day in off-plant meetings. An outcome of this approach was that personnel making PCBs were asked what was stopping achievement. They came up with the answer 'cut out queues'. While they did not know how to achieve this they were able to identify that the problem lay between the production plan, the kitting area and the production process. This pointed to areas for development.

Detailed strategies were developed for each area.

People involvement   This strategy had four key goals.

1. Cycle time reduction.
2. Effective inventory management.
3. On-time production/shipments.
4. Climate for continuous improvement.

To achieve these goals a programme was commenced of educating managers and supervisors in teamworking. Within nine months it was realised that the four key goals outlined above were not the objective of people involvement.

People involvement was the way in which the goals of total quality strategy, MRP/JIT and loss prevention could be achieved.

Supplier partnering did not receive the same emphasis as the other three areas because purchasing considered that, until Fisher Controls gained experience in controlling its own manufacturing process, it would be difficult to get the message across to suppliers.

To prototype the idea of teamworking, a target group was selected to train and organise as a self-managing work team.

A programme was instituted to identify likely training needs, skill or technically based, social or interpersonal skills.

Steps were taken to identify and decide upon likely 'blockers' in pay and grading systems, health and safety, and formal administrative systems.

Total quality strategy

Approvals.

- 1990– maintain British Standard BS 5750/ISO 9000 approvals.
- Future – aim for zero audit deficiencies, 50 per cent reduction by 1991; standard worldwide quality approvals by 1992.

Quality improvement programme.

- 1990 – complete quality education for all employees; commence follow up education (i.e. ten steps); zero defects programme.
- Future – improvement programme to be supported by a worldwide zero defects programme for marketing through design, manufacturing, project management, delivery, install/commissioning and after care; to be no. 1 in customer satisfaction; to carry out worldwide quality education.

Quality organisation.

- 1990 – re-structure for improved communication, focus, customer planning, inspection methods/tools.
- Future – remove traditional inspection within three years and replace by: supplier partnering, vendor quality plans, source/skip inspections, patrol/audit inspections.

Quality plans.

- 1990 – local and worldwide to co-ordinate improvement activities.
- Future – deliver superior quality goods and services achieving quality leadership; 50 per cent reduction in customer failures; 50 per cent reduction in production failures; quality plans generated, owned and actioned by each department.

Supplier partnering

Reduced supplier base.

- 1990 – achieve approximately 190 stock suppliers; preferred vendor list to be in operation with technology and marketing 'signed off' on its application; partnering potential to be identified and agreed; initial formal contracts to be made with most critical suppliers.
- Future – supplier base around 50, with 80 per cent partnering to a significant degree and a strict control on the addition of new vendors.

A shared MPS (master production schedule) with suppliers.

- 1990 – leadtime distribution to allow a greater match between MPS and supply; vendor scheduling introduction to provide more stable supplies and critical unit assembly material; improved stability to allow de-expediting to become a routine; pilot vendors for shared MPS to be selected and trials started.
- Future – critical suppliers to have shared MPS via electronic data interchange (EDI) communication, then sign-off on plan changes within forty-eight hours; inventory holdings of one week or less; expediting to be a rare event.

Leadtime reductions.

- 1990 – move towards target effective purchased leadtimes of three months; Establish alternative order policies for items in excess of three months.

Case Study 2: Fisher Controls Ltd

- Future – purchased parts to have a maximum leadtime of six weeks shared design/development/phase-in/phase-out.
- 1990 – purchasing to develop closer links with marketing and technology; greater use of preferred vendor list; new vendors to be formally approved prior to being designed in.
- Future – Fisher will have access to knowledge where partners freely share their R&D plans; suppliers are actively involved in new product development. Phase-in/phase-out have no inventory impact.

MRP/JIT

On-time production/shipments.

- 1990 – 90 per cent to the master production schedule (MPS); 90 per cent delivery integrity; focused MPS/capacity planning.
- Future – one hundred per cent to the MPS; 98 per cent delivery integrity to customers' request; 100 per cent delivery integrity to promise.

Inventory management.

- 1990 – achieve/beat budget (£8.5m) inventory; manage worldwide PCB excess.
- Future – worldwide inventory management; no-stores goal – line inventory only; Leicester inventory value of £2m; manufacturing – built stock of zero (no stock rework).

Total stacked leadtime.

- 1990 – average throughput time on MPS items to be three weeks or less; twelve weeks maximum leadtime for purchased parts.
- Future – one day turnround (kit to build); build for sales orders only (no stock) no shop reworks; six weeks (or less) maximum leadtime for purchased parts.

Loss prevention

To be the safest company there is with zero accidents.

- 1990 – not more than two recordables and sixty first aid injuries; loss prevention audit (LPA) score of twenty-three or less with broad, general awareness.
- Future – a business where unsafe acts and conditions do not occur, where management knows first if our products, processes or people are likely to cause potential problems.

Zero environmental impact.

- 1990 – waste minimisation plan agreed and quantified for hazardous waste; legal requirements to be met.
- Future – a business with recognised and demonstrated concern for our neighbours, our employees and both local and worldwide environments; a business which maintains a continuing waste minimisation programme (for hazardous and non-hazardous wastes) which emphasises source

reductions and recycling with the ultimate goal of achieving 50 per cent or less total waste (consumption).

Zero losses.

- 1990 – major losses to be recorded and investigated.
- Future – a company which is aware of the value of its assets, the potential for loss and which has implemented control measures to eliminate such losses.

CS2.5 Process development

The manufacturing strategy had a direct impact on the production process requiring a change to product layout if the goals of the manufacturing strategy were to be realised.

CS2.5.1 Layout of production and changes made

Originally, layout was by process, planning produced weekly schedules and each department carried out its process and transferred the partly completed board to the next process as shown in Figure CS2.1. The production flow shown in Figure CS2.1 was disconnected. Each process was run by a group with specialist skills, managed by traditional supervisors and supported by a number of manufacturing engineers with no common understanding or goals.

Towards the end of 1990 the problems of this layout were identified. They are shown in Figure CS2.2.

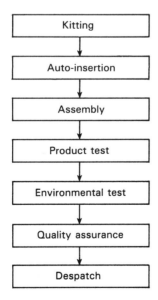

Figure CS2.1 Process layout – disconnected flow

Case Study 2: Fisher Controls Ltd

> Queues
> Multiple handling
> Pass the problem
> Difficult to balance flow
> No true ownership
> Incompatibility with JIT philosophies

Figure CS2.2 Problems of disconnected flow layout

A decision was made to reorganise production. The step change that was needed was to move to a product line structure, sometimes referred to as cells, teams or a factory within a factory. Each product line would make a family of products and each product line would have a strong emphasis on team working aimed at a multi-skilled approach achieved through continuous training. Product lines would all have a product flow as shown in Figure CS2.3.

Figure CS2.3 shows a connected flow with all activities carried out by a team of multi-skilled workers using dedicated equipment (with the exception of the auto insertion machines and the flow solder machine). Quality assurance technicians became line auditors working with each team to ensure the use of up-to-date process procedures.

Each line had a team leader with clear objectives. These included:

- Advising the planning department on cell capacity.
- All material required by each line was to be the responsibility of that line.

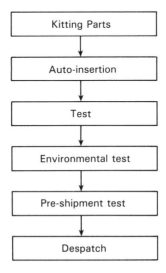

Figure CS2.3 Product flow

Process development

This would ensure that the main computer system could be continuously updated so that a valid production schedule could be produced without any shortages.
- Quality was to be the responsibility of the line. The line would have ownership of quality and would be measured by achievement of quality.
- Cycle time would be the time the product line took to produce high quality product aiming for zero defects, i.e. the emphasis was to be on quality not volume.
- On-time production was to be measured against promised delivery date to customer.

Resistance to change

Changing production strategy was not easy. Uncertainty of the outcome of the changes caused anxiety. The entire manufacturing function did not react favourably. Some saw the change as a threat to their authority by allowing the shop floor to make key decisions about such things as process ownership, equipment ownership and production support.

The resistance to change can be illustrated by the following:

1. Staff were worried that the changes would not be suitable to their career development. In fact, only 1 out of 110 production staff left for this reason.
2. Reluctance to change away from old working practices. This was particularly pronounced in the case of manufacturing engineers.

Problems also arose from the change. If cell boundaries were rigidly defined there would be a lack of co-operation between the production teams. However, fluid boundaries could lead to problems. For example, the requirement for hi-pot testing was not fully documented with the result that a cell failed to have the test carried out and the product did not carry a test stamp. This was unacceptable in USA and when the product arrived it was rejected.

However, Table CS2.1 shows the benefits accruing from the adoption of the manufacturing strategy and the subsequent move from process focus to product focus.

Table CS2.1 Benefits from the manufacturing strategy

	1988	1989	1990	1991	1992
Quality yield	70%	85%	90%	94%	95%
Cycle time (weeks)	14	10	4	1.5	3 days
Delivery time (Required v promised)	50%	60%	78%	95%	97%
Inventory	£12m	£9.8m	£7.9m	£3.7m	£2.5m
Vendor base		560	456	159	154
Non-conforming material receipt (No. of items)	203	184	65	5	5

Case Study 2: Fisher Controls Ltd

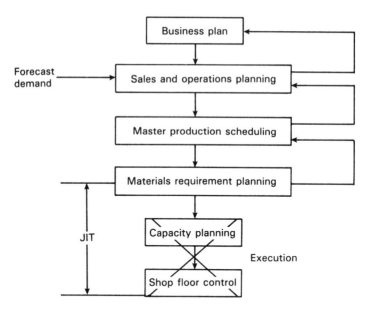

Figure CS2.4 Modification of MRPII

CS2.6 Improvements effected by the strategies

The MRP/JIT strategy contributed to the improvements shown in Table CS2.1. JIT also resulted in change in the execution modules of MRPII. These were replaced by JIT. This allowed concentration on the planning and provisioning aspects, as shown in Figure CS2.4.

CASE STUDY 3

Hepworth Building Products

CS3.1 The justification for manufacturing and selling concrete products

CS3.1.1 The manufacture of concrete pipes, manholes and gullies

The expectation was to obtain 20 per cent of the expanding concrete sector within 5 years, using existing distribution channels and current drainage sales force.

To achieve this an automated factory was to be built on a greenfield site to produce 300,000 tonnes per annum.

Key data

Capital cost	£14.7m
Sales revenue per annum	£21.9m (after 5 years)
Profit per annum	Very good
DCF (after tax)	In excess of 20 per cent
Payback	5 years

CS3.1.2 Strategic considerations

Hepworth manufactured clay and plastic drainage pipes. Entry into the large diameter concrete sector, worth £90m at ex-factory prices, was not possible since clay incurs substantial cost penalties in larger sizes. Clay is only competitive up to 300 mm.

Entry into the concrete sector would strengthen the market position of the company and, by utilising existing distribution channels, would allow Hepworth to compete for all elements of multi-material drainage contracts (concrete, clay and plastic). No other supplier had the possibility to become multi-material.

The main markets for concrete are sewerage and roads. Both markets were depressed between 1979–1986 because of cutbacks in capital expenditure. Government policy changed in favour of infrastructure investment and prospects for the 1990s were good. Hepworth intended to take 20 per cent of the concrete sector and this share was expected to be fully absorbed by anticipated growth in sector volume, diminishing the risk of competitor retaliation.

CS3.1.3 Entry into concrete

Hepworth's range of concrete pipes was to be the best on the market, manufactured on state-of-the-art equipment, incorporating a high quality jointing seal giving an outstanding end user benefit.

Case Study 3: Hepworth Building Products

Hepworth would be the lowest-cost producer in the UK and would be price competitive despite high initial depreciation charges.

The product would be exploited by a drainage sales force of ninety compared to the market leader who only employed seven.

Hepworth, through its proven contract reference system, already called on every specifier and contractor involved with sewerage and roads. The Hepworth clay drainage manual was accorded the status of a standard reference.

Leading heavyside merchants who supplied 50 per cent of this market had already expressed strong support for Hepworth's entry and were prepared to displace existing suppliers.

Hepworth's concrete pipe plant by virtue of its size and automation would enjoy substantial scale economies.

The location at Ellistown, a greenfield site, was ideally positioned in the centre of the UK with excellent motorway access and was near Hepworth's transport complex at Woodville. In addition there were excellent supplies of raw materials in the locality, cement at Rugby, granite within one mile and sand from Trent valley.

Concrete technology was not inherently difficult and proven equipment was readily available from a variety of continental suppliers. However, it was the aggregation of all the latest technology and full automation of the plant that would ensure sustainable cost advantage.

A strategy of acquisition was rejected on the grounds of cash flow and vulnerability to a new entrant using greenfield entry and state-of-the-art investment. RMC or Tarmac, with major involvement in road construction, could have developed a new entrant strategy.

A second alternative, the development of a cost-effective large diameter clay pipe, had been extensively explored. The clay range could go up to 800 mm diameter and there was, therefore, part of the proposed range which could be either clay or concrete. Clay could not be developed above 800 mm diameter and it was decided that the new concrete facility should cover the pipe range from 300 mm to 1800 mm, although clay pipes up to 800 mm will continue to be made by existing tunnel kiln methods.

CS3.2 Hepworth entry strategy

For Hepworth concrete and clay were parallel markets. The company already tendered for the clay component of most contracts. Distribution methods were identical and the contractors and merchants were largely existing Hepworth customers. Apart from the current market leader competitors were small and fragmented.

CS3.2.1 Key elements of Hepworth's proposition to the market

- Product range. Hepworth would offer a full range of improved concrete road and sewerage products. These would comprise:

    – flexibly jointed pipes 300 mm to 1800 mm, incorporating, for the first time in the UK, a captive seal for improved jointing. There were significant problems in the trench when jointing with the existing loose ring

Hepworth entry strategy

system. The Hepworth product would offer a significant advantage to the contractor in terms of assembly time and security from leakage;
- manhole rings and cover slabs 675 mm to 2700 mm incorporating improved methods of lifting;
- road gullies 375 × 750 mm deep to 450 × 900 mm deep incorporating adaptable clay or plastic connections.

- Assured manufacturing quality. All products would be manufactured to BS 5911 and kite marked accordingly. This in itself was not unique but Hepworth pipes would be 100 per cent pressure tested for porosity during manufacture. No other UK manufacturer had this ability and would find it very costly to follow Hepworth's lead.
- Delivered quality. Smaller diameter pipes would be palletised for easier storage and handling thereby reducing risk of damage. Mechanical off-loading facilities would be offered as standard to minimise risk of damage and assist the contractor to position the pipes on site. No other manufacturer offered this facility which was derived from Hepworth's clay pipe experience.
- Hepworth sales force and reputation. Hepworth had a drainage sales force of ninety, more than three times the combined total of the concrete industry. Hepworth had undisputed credibility as a leading drainage supplier in terms of product quality, technical excellence and delivery reliability.
- Multi-material product range. With its range of clay, plastic and concrete Hepworth would be unique and very difficult to ignore.
- Ex-stock availability. Small frequent-use items would be stocked at depots in the same manner as clay. Contractors and merchants would gain the confidence of the local top-up facility currently enjoyed with clay.
- Merchant-only policy. All sales would be routed via Hepworth merchants. This would reinforce the strategic objectives of holding clay sales and developing plastic and bricks by providing a total range of drainage products.

  Approximately 50 per cent of the market was routed via merchants. Hepworth's key task would be to displace existing suppliers in Hepworth merchant outlets.

  When competitors were displaced and supplied the market direct they would enjoy a potential 5 per cent price advantage but this was fully taken into account in Hepworth's price intentions. In essence Hepworth would still be price competitive despite the 5 per cent discount and this was allowed for in the financial appraisal.

  The long-term benefits to clay, plastic and brick, manufactured by other group companies, were significant but not quantified in this proposal as the case was considered sufficiently robust in itself.
- Promotion. Launch costs would be heavy in year one to establish a presence. With a tightly defined target audience, direct promotion in terms of launch conferences, direct mail and technical handbooks would be most appropriate. Thereafter, the level of expenditure would fall significantly as concrete became part of the main building products range.
- Transport economies. The location of the plant would take full advantage

Case Study 3: Hepworth Building Products

of the Woodville transport complex nearby and would be able to reach, economically, 85 per cent of the proposed new road building programme.
- Price position. Hepworth intended to be price competitive with the market leader.

CS3.3 Manufacturing

CS3.3.1 Production processes

The first flexibly jointed pipes for gravity sewers and storm water drains were produced by the spinning process. In this method the concrete was fed into a horizontally spun mould and steam cured before the pipe was demoulded. This was a slow and labour-intensive process.

In the 1960s the packerhead machine supplied by McCracken and Hydrolite arrived from the USA. The machines compacted pipes vertically, very quickly and with very dry concrete mix. The pipes had obvious cost benefits but unfortunately did not reach performance requirements, particularly regarding hydrostatic properties, due to lack of consolidation of the concrete.

The problem was overcome by the introduction of the vibration technique which was developed originally by the Danish company Pedershaab. In this process pre-mixed concrete was conveyed into a mould forming the outside profile of the pipe. The internal dimensions were formed by the insertion of a mechanically vibrated core which consolidated the material to give the required compaction. The mould and core were then immediately removed for the next cycle and the cast pipe transferred to a curing chamber for a period of 24 hours.

The vibrated pipe principle would form the basis of manufacture in the new factory.

CS3.3.2 Pipe joints

Flexible joints were originally achieved by a rolling 'O' ring which often became misplaced in the jointing process causing leakage to occur. An alternative ring was introduced with a 'G' or pear-shaped configuration. This was adopted by the market leader and became widely used, particularly in smaller diameters.

It was, however, possible for the ring to be misaligned and whilst many site difficulties were overcome by the 'G' ring it was not a 'captive ring' and, therefore, exact location could not be guaranteed. Continental Europe developed a captive seal (Glipp) located in the socket of the pipe. This was considered to be the most effective joint available, and was incorporated into the Hepworth product range. Jointing is achieved by greasing the pipe end and pushing it into the spigot of the adjoining pipe similar to the method of jointing plastic plumbing.

The Forsheda 'Glipp' captive seal was widely used in continental Europe but had not been introduced in the UK because of tooling costs. Any existing manufacturer wishing to use this joint would have to replace all their pallet rings. Trade and specifier reactions to the Glipp seal were favourable. In June 1991 the managing director of the new plant said that a contractor was achieving 50 per cent faster laying because of this seal.

CS3.3.3 The new plant layout (Figure CS3.1)

The raw materials, crushed rock, sand, cement and filler are transported to site by lorry and stored in large overhead silos adjacent to the batching plant. The mixing equipment, manufactured by Skako, has the following features:

- Automatic moisture control.
- Accurate weighing of raw materials, leading to greater material control.
- Fast and effective distribution of concrete to the pipe-forming machines by skip on an overhead mono-rail.

Batched concrete is conveyed to the following machines:

- Machine 1 – pipes 300 mm to 600 mm diameter. Two pipes are cast simultaneously on the machine and then transferred to transport cars for processing through a curing chamber. Curing chambers are draughtproof tunnels at ambient temperature. Draughts cause surface cracking. After a minimum of 12 hours the concrete is sufficiently hard for the pipes to be automatically removed and placed on a line, hydrostatically tested, stamped and despatched to the stockyard where they are cured for 10 days to achieve full strength.
- Machine 2 – pipes 675 mm to 1200 mm diameter. The pipes are cast on the machine individually and then processed as above.
- Machine 3 – pipes 1350 mm to 1800 mm diameter and manholes 1800 mm to 2700 mm diameter. This is a double station machine, one for pipe and one for manholes. The product casting sequence is automated but because of the low production rates the product and its mould are removed manually and transported to the curing hall by overhead crane.
- Machine 4 – manholes 900 mm to 1500 mm diameter. This is a dedicated machine and is the most automated method of production. Can be used to make a range of additional flat concrete products.
- Machine 5 – manhole covers 900 mm to 1200 mm diameter. The sizes of these products preclude them from being made on machine 4 and they will, therefore, be produced by a simple vibrating table and mould system.
- Machine 6 – manhole covers 1350 mm to 2700 mm diameter and road gullies 375 mm and 450 mm diameter. These are cast automatically on the machine and transported to the curing hall by a small electric cart.

CS3.3.4 Reinforcement cage machine

A machine for on-site manufacture of reinforcement cages for pipe diameters above 600 mm has been installed.

CS3.3.5 General

All machines have been selected to meet production at full output on a double shift, 7-day week.

Case Study 3: Hepworth Building Products

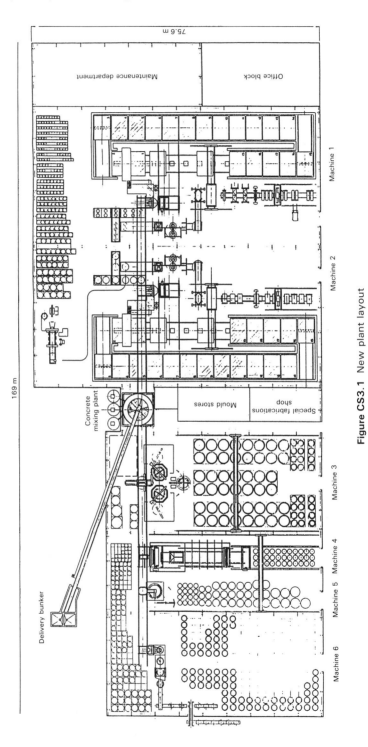

Figure CS3.1 New plant layout

Software constraints

CS3.3.6 Sequence of operations for pipe making

The pipe is formed on a base ring that allows a robot to lift the pipe vertically as soon as forming is completed and transport it to a pre-selected position on a moving table. The Glipp seal has been fitted into the socket during the forming operation.

When full, the table moves into the curing chamber for that line and in 12 to 24 hours the pipe is picked up by a robot. The pipe is now sufficiently strong for the robot to grasp the pipe and transport it to a machine that extracts the support ring. The robot turns the pipe through 180 degrees and places it onto a horizontal track for inspection, testing and labelling.

CS3.3.7 Quality control and testing

All products are visually inspected and stencilled with the company name and product details. Pipes up to 1200 mm are subject to an online hydrostatic test, above this size the testing is in batches. Product that fails to meet standard is either scrapped or the fault rectified.

CS3.4 CIM strategy

A strategy was implemented covering the following areas of the business:

- Material requirements planning including capacity planning and work scheduling.
- Inventory control.
- Sales order processing (SOP).
- Plant operations.
- Management reporting.

CS3.5 Hardware constraints

Production plant had already been ordered, therefore, it was not possible to specify the requirements for input and output to the controlling programmable logic controllers (PLCs). In addition there had been no choice of PLC manufacturer so a consistent plant interface could not be guaranteed.

The business systems ran on IBM System 38 and AS400 computers. As some of these systems were to be retained any proposed new systems had to be able to communicate with the required level of interactiveness with these computers.

CS3.6 Software constraints

Existing systems were to be used where possible. This affected:

- SOP systems that had been developed by Hepworth to cater for current business needs.
- The company used an MRP package supplied by JBA on a number of sites. This product was to be utilised in order to provide consistency.

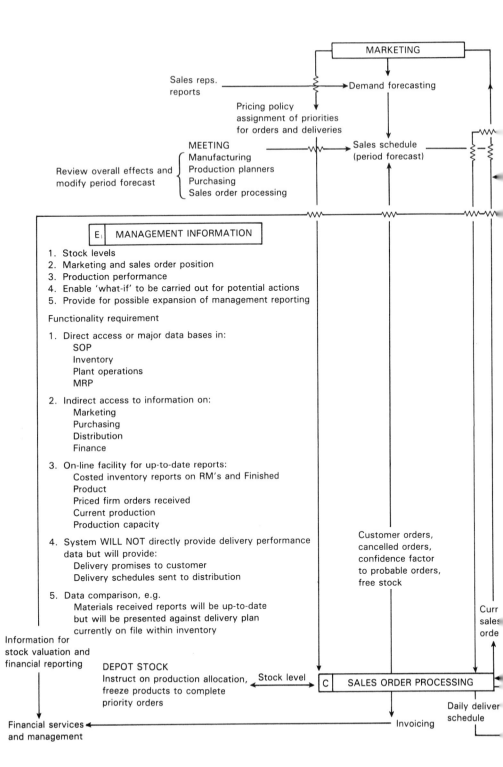

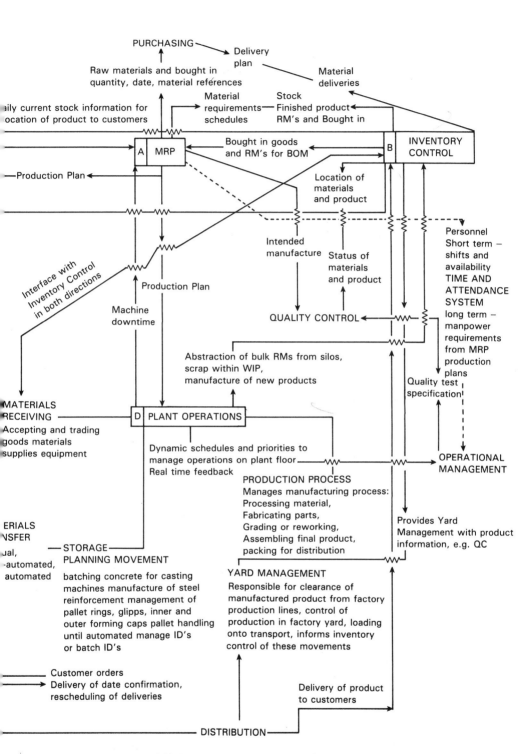

Figure CS3.2 Relationships within the CIM system

Case Study 3: Hepworth Building Products

CS3.7 Description of the CIM system

The relationships within the CIM system, analysed in this section, are shown diagrammatically in Figure CS3.2. The five areas are:

A. MRP – Section CS3.7.1.
B. Inventory control – Section CS3.7.2.
C. Sales order processing – Section CS3.7.3.
D. Plant operations – Section CS3.7.4.
E. Management information – Section CS3.7.5.

CS3.7.1 Materials requirement planning

MRP was viewed as a module of MRPII interfacing to inventory control and sales order processing. Although production planning will initially be governed by the 'make to stock' parameters, there is the longer-term need to reduce inventory and respond more quickly to both customer and market demands. This requires a move towards a 'make to order philosophy'. The MRP system must satisfy both of these operating policies concurrently.

Functional areas

- Maintenance of production.
- Constraint data.
- Production scheduling.
- Capacity scheduling.
- Purchase requirements.

These can be used to reduce costs and to increase inter-functional liaison within the company to form an effective business strategy.

Interfaces with MRP

- Management. Data must be presented to management in a way that allows executive decision taking across company functions.
- Personnel. For short-term planning the skills and availability of manpower can be obtained from a time and attendance system. To meet long-term policies, manpower requirements can be obtained from MRP production plans.
- Marketing. A cycle of demand forecasting operates as follows: marketing assesses probable demand for each product from external company statistics and sales representative reports. This demand is entered in to the sales schedule to produce a period forecast. The 'what-if' facility should be explored to confirm the effect of marketing activities. A meeting is then held with manufacturing, production planning, purchasing, SOP and inventory control to decide the overall effect on the company of the period forecast. The period forecast is modified. The cycle is repeated.
- Sales order processing. Marketing provides the period forecast. SOP

Description of the CIM system

updates this forecast by the inclusion of firm orders and cancelled orders that will alter manufacturing and purchasing requirements. There is a facility to allocate confidence factors to probable orders which can then be included in the updated forecast. In the short term for 'make-to-stock' this is low priority but becomes important as the switch is for 'make-to-order'.
- Inventory control. Data required by MRP: free stock in order to calculate manufacturing requirements in the production schedule; bought-in goods in order to itemise bill of material lists and the routings utilised in production planning; raw materials to ensure that factory scheduling will reflect a true situation of current circumstances on the shop floor.
- Purchasing. Notification of the requirements for raw materials and bought-in goods consisting of material references, quantity and date needed.
- Quality control. An online facility to notify intended manufacture to QC and relate the result of QC checks to inventory control.
- Factory supervisor. Machine downtime must be notified to MRP.

CS3.7.2 *Inventory control*

The maintenance of low stock levels means that the planned production process is vulnerable to errors in materials handling or divergences from anticipated materials delivery and usage. As a result inventory control must be integrated with other important areas such as plant operations, MRP and purchasing to monitor and report discrepancies in materials received and issued.

Functional areas

- Record and report details of material deliveries. A delivery plan produced by purchasing from the MRP schedule is needed for comparison with actual deliveries.
- Storage of delivered materials. To ensure the traceability of materials, each delivery to store should be accompanied by unique identification of the materials delivered.
- The issue of materials from store. Authority levels must be set for issue requests and issues must be made to authorised individuals. Plant operations will provide information about the consumption of bulk materials from silos.
- Preparation of movement histories. The outputs are the rate of materials movements for defined periods and the actual quantities of materials in store at any point in time.
- Updating recorded stock levels. Stocktaking and reconciliation of the actual materials against records to control sources of error such as underdelivery of materials, shrinkage and unreported withdrawals from stores.
- Preparation of reports. Management will require periodic reports on materials and product quantities, by individual material or product, or by group. Comparison of usage with turnover will be needed.
- Updating of recorded product balances. The product inventory will be

Case Study 3: Hepworth Building Products

updated from information on product movement supplied by yard management, distribution, depots and QC.

Interfaces

- Financial services. Will use information supplied by inventory control to value stock and prepare financial reports to management.
- Purchasing. Purchasing control the ordering of materials and finished goods in accordance with directions from management and MRP. Purchasing will inform inventory control of anticipated deliveries from suppliers and will receive details of actual receipts from the inventory system.
- Quality control. For QC to monitor the acceptability of materials and product within the materials store and factory yard, information is needed from inventory control on the location of materials. QC will input information on the status of the materials and product to update the inventory.
- Distribution. Will deliver product in accordance with the schedule produced by SOP but will inform inventory control of the delivery of product to customers.
- Plant operations. Will provide inventory control with the abstraction of bulk raw materials from the silos, scrap within work-in-progress and the manufacture of new products.
- MRP. Will interface with the inventory system to determine the orders to be placed with suppliers and the products to be manufactured. Will receive information on actual stock levels from inventory control for manufacturing planning. Will inform inventory control of the anticipated materials requirements scheduled. Inventory control will use this information to monitor actual consumption of materials.
- Yard management. Responsible for the clearance of manufactured product from the factory production lines, the control of all product within the factory yard and its loading onto delivery transport. It informs inventory control of these movements. Inventory control receives information from other sources, such as QC, and provide yard management with information on the location of product within the yard.
- SOP. Inventory controls provide SOP with current product stock and its status to enable SOP to allocate product to customers.

CS3.7.3 Sales order processing

Functional areas

- Service and schedule orders. Monitors customer orders against planned delivery. Gives delivery date confirmation to clients and re-schedules deliveries as necessary. Answers enquiries from customers and merchants about scheduled deliveries. Prepares invoices when the goods are delivered to the customer. Information is supplied to financial services and management as necessary.
- Control depot stock. Receives and monitors depot stock level informa-

Description of the CIM system

tion. Instructs the depots on product allocation and will 'freeze' products at the depots if necessary to complete priority orders.
- Receive merchant customer orders. SOP provides a forecast delivery date for a potential order resulting from an enquiry by a customer or merchant. It records and acknowledges the order.
- Schedule orders and capacity. Maintains information within SOP which is vital for the functioning of the other processes.

Interfaces

- Financial services. Financial services will provide financial control to all areas of the Hepworth organisation. SOP will receive priority information from marketing and will inform financial services of invoicing details.
- Inventory control. SOP will receive daily current stock information from inventory control for the allocation of product to customers.
- MRP. SOP will receive daily updates of the manufacturing production plan from MRP in order to schedule product delivery promises to customers. It will inform MRP daily of the current sales orders status.
- Depots. SOP will direct the depots on the allocation of products.
- Distribution. SOP will provide daily schedules and forecasts of transport requirements. Actual performance will be fed back.
- Marketing. Marketing is responsible for the management and direction of the long-term sales strategy. Marketing will inform SOP of the pricing policy to be adopted with customers and merchants and the relevant discounts. It will direct SOP on priority assignments for orders and deliveries. SOP will inform marketing of the current sales position.

CS3.7.4 Plant operations

Functional areas

- Production management. Production management provides dynamic scheduling functions for the plant floor by assigning priorities, personnel and machines. Input for production management includes new orders and schedules from the MRP production plan. Output includes dynamic schedules and priorities used to manage operations on the plant floor and real-time feedback from plant operations.
- Materials receiving. This function includes accepting and tracking goods, materials, supplies and equipment from outside suppliers or other locations within the enterprise. Materials receiving will use as its major input the data obtained by the various sensors within the plant, the numerous PLCs, the automated processes and any data entered manually via operator workstations. The boundary of the material receiving function will be the automated process that assists delivery of the raw materials, i.e. sand, aggregates and cement to the storage silos. Receipt of raw materials will be recorded and materials receiving will have strong interfaces with inventory control.
- Storage. Storage is responsible for the planning of movements such as the batching of the raw materials for delivery of concrete to the casting

Case Study 3: Hepworth Building Products

machines. Other plans include those for the manufacture of steel reinforcement structures for certain concrete products and also for the management of items such as pallet rings, Glipps and inner and outer forming caps (which maintain the inner and outer dimensions of the pipes for a period following the initial casting process). The pallet handling and management requirements may be automated. Initially the process will be managed by plant operations. This function will handle the item IDs or batch IDs that will be assigned to materials, work-in-progress and finished goods, against which all data will be recorded and tracking facilities developed.

- Production processes. The production process function manages the manufacturing process on the plant floor, including processing materials, fabricating parts, grading or reworking, assembling final products and packing for distribution.

  The data flows from this function are as follows:

  – PLC parameters/programs.
  – Operator schedule.
  – Plant instructions.
  – Plant information.
  – Performance statistics.
  – Machine performance.
  – Product raw materials usage.

  The data flows into this function are as follows:

  – Plant data.
  – Batch IDs.
  – Stoppage reasons.
  – Events records explanations.
  – Manufacturing parameters.
  – Scrap values/causes.

- Quality test and inspection. Each individual product will be checked both manually and using automated plant. Data will be gathered by production batch rather than by specific product. The quality test specification will be input by operational management. The data will be output to operational management, engineering and inventory control as part of the data associated with a particular production batch.

- Material transfer. This function handles the movement of all materials round the plant and has a significant interface with the material storage function. The activities within this function may be manual, semi-automated using control panels, fork-lift trucks and conveyors, or fully automated relying on PLCs, distributed control systems, stacker cranes, programmed conveyors, automated guided vehicles and pipelines. The inputs to and outputs from this function are automated and non-automated. For the latter entry will be made at a workstation.

- Plant maintenance. This functional area can have a major impact on the plant's overall performance. There are three broad classes of activity:

  – Preventive maintenance.

Description of the CIM system

- Emergency maintenance.
- Inspection and overhaul.

- Plant site services. The management of all utilities, environmental control and other such services is an essential part of plant operations. The data will be obtained from sensing devices and measuring equipment, e.g. factory temperature and humidity readings which may affect the manufacturing process.

Interfaces

- Factory supervisor. Represents the operational management personnel directly responsible for the operation of the plant who will make the necessary manual decisions that will control functions within plant operations. Factory supervisor will undertake the monitoring of the information supplied by plant operations and adjust manufacturing and working parameters to obtain maximum efficiency and optimise output in accordance with the MRP factory schedule.
- Inventory control. Inventory control functions will undertake recording, storage and monitoring of all inventory items. They will also manage all consumable items. Inventory control will inform plant operations about deliveries of raw materials that require physical movements or a change of state to be actioned by the plant which is controlled by plant operations. Plant operations will provide the required information about inventory within the plant and/or under the control of plant operations. This information will include confirmation of raw material delivery, raw materials consumption and movements of products within the plant.
- Maintenance/engineering. Responsible for taking the necessary action to ensure the continued operation of the production process and other facilities within the factory. Information concerning the quality of the manufactured products will be supplied in order to assist engineering in assessing the impact of changes to the plant as well as its day-to-day operation and performance.
- Management. Operational and business management who require information about activities controlled by plant operations will determine the exact nature of the information required.
- MRP. The MRP function will provide the necessary manufacturing requirements plan for a period from current day to a future horizon, e.g. one week. This will be passed to plant operations as the factory schedule that will be used to undertake any appropriate planning within the plant operations environment and also generate the manufacturing plan and operator schedule from which activities within plant operations will be driven.
- Plant. Plant refers to all automated plant and devices within the factory that are controlled and/or monitored by plant operations. This will include mechanical, electronic and computerised equipment such as PLCs.
- Plant operator. Plant operator refers to any non-automated function within plant operations. These may be operators at a machine, quality checkers, supervisors, maintenance engineers or other personnel who

209

Case Study 3: Hepworth Building Products

are authorised to have access to data or instructions or who may need to input data that is not obtainable by other means.

CS3.7.5 Management information

The importance of management information within an integrated MRPII system is not so much with the day-to-day control of the various functions of the system which are covered by the rules of the functional specification. Rather it is with the longer-term identification of trends, the direction of strategy and the resolution of problems. It is vital that the information presented to management is comprehensive, accurate and consistent across the whole system in content and timing. It is also vital that this information is available to the right personnel at the time it is required for the appropriate actions to be taken.

This can only be achieved by an integrated corporate database supported by a full range of reporting and enquiry facilities.

Objectives

- To provide accurate and timely structured reports to management on:
  - The stock levels in all areas of production and finished product.
  - The current marketing and sales order position.
  - Production performance.
- To enable management to access information rapidly at all levels and to be able to carry out modelling enquiries ('what-if') on the overall consequences, for the areas covered in the functional specification, of potential actions.
- To provide ad hoc reports and comparisons as required.
- To enable the management reporting facilities to be expanded at a later date to encompass other business areas and reporting enhancements.

Functionality requirements

The system will:

- Directly access the major databases in the functional areas of MRP, inventory control, SOP and plant operations.
- Indirectly access information on marketing, purchasing, distribution and finance.
- Provide a continuous online facility for up-to-date report generation of costed inventory reports on raw materials and finished product, on priced firm orders received, and for current production and production capacity.

The system will not be able directly to provide delivery performance data, but will be able to provide details of delivery promises made to customers and delivery schedules sent to distribution.

Within the various reports, the system will provide online information where possible and include relevant data for comparisons. For example, the materials received reports will be up-to-date but will be presented against the delivery plan that is currently on file within inventory.

CASE STUDY 4

Kodak Limited

CS4.1 Introduction

Kodak at Annesley in Nottinghamshire is part of the large American Eastman Kodak company. This case study shows successful class A MRPII implementation as a key element in the plant's TQM strategy. Class A is reached when the MRP system provides the company plan which the management use to run the business (see Appendix 1 of this case study on p. 218–19 for class A definition).

The case study concentrates on the manufacturing element of the Kodak 35 mm supply chain. The distribution and marketing implementation phases have since followed completing the MRPII supply chain integration from suppliers and subcontractors through the manufacturing process to the final customer. Significant business benefits have resulted.

JIT techniques were introduced at shop-floor level as part of the process, excluding detailed shop-floor control from the MRPII computer system. Formality, integration and simplicity were bywords in this successful class A implementation.

The strategic use of MRPII has been developed further by concentration on 'time fences' which are discussed in Appendix 2 of this case study on p. 220. The time fences provide guidelines for every product which demarcate when change is feasible and, beyond which, change cannot be made.

CS4.2 The manufacturing process

The manufacturing process slits and perforates film which is then put into cassettes. The cassettes are put into cartons and in turn these are made up into bulk packs ready for despatch. There are two manufacturing lines:

- 135 films operating at volumes of hundreds of millions p.a.
- Roll films operating at volumes of tens of millions p.a.

Figure CS4.1 shows the flow of products through the manufacturing operation. The manufacturing processes are:

- The plastic operation where moulding machines form spools, lids and containers.
- Cassette manufacture by cutting and forming metal and producing an end cap. The cassette has to be lined with velvet where the film exits to provide a lightproof seal. The end cap has to fit onto the cassette and locate the spool to form a lightproof seal. Precision has to be high in order to allow free rotation of the spool as the film is unwound and rewound.
- Slitting and perforating film.

Case Study 4: Kodak Ltd

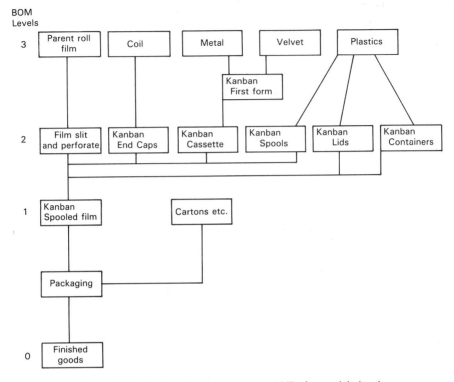

Figure CS4.1 Product flow, kanbans and bill of materials levels

- Spooling the film into cassettes.
- Packing the cassettes into plastic containers which are placed in purchased printed cartons.
- As stated the cartons are finally put into packs and palletised ready for despatch.

Tolerances are tight for the formation of spools and cassettes to ensure correct camera function. Tolerances are also tight to satisfy the process needs of high speed spooling. Similarly, tolerances for packaging materials must be set and maintained so that high speed packaging lines will work effectively.

CS4.3 Bill of materials (BOM)

Figure CS4.1 also shows the four levels in the BOM which controls 250 finished products and 1,500 items. Originally there were five levels but one level was eliminated by the use of kanbans.

CS4.4 MRPII and change

For the previous nine years a predecessor to MRPII had been in use in the company. This was SAM (System for Annesley Manufacturing) which was a

Resource commitment

computerised, non-order-based production and inventory control system. It was realised that changing from this system to MRPII was a major step and to combine this with the introduction of kanban as the means of shop-floor control was going to bring about even bigger changes.

An aggressive implementation schedule was adopted of between eighteen months and two years. The short time scale was adopted to:

- Generate intensity and enthusiasm.
- Focus on MRPII/kanban as the priority.
- Regard MRPII/kanban as the vehicle driving all the changes which would be needed to achieve success.
- Prevent schedule slippage.
- Realise benefits quickly.

CS4.5 The plan of implementation

In December 1988 a steering team was formed which held an off-site workshop at which it was agreed to:

- Adopt the Oliver Wight Associates' 'proven path' as the general implementation approach, see Figure CS4.2.
- Form a commissioning team.
- Provide an education programme for the management team.

Andy Coldrick of the Oliver Wight Associates was chosen as the Kodak Ltd MRPII consultant. Implementation was seen as a project with three primary variables:

1. The amount of work to be done.
2. The amount of time available.
3. The amount of resource available to accomplish the work.

Initially the workload and the time were considered constant. The only variable was seen to be the resource. This resource had to be withdrawn from the production team and, at the same time, production had to be maintained. Subsequently the workload became greater than anticipated and the project was extended. However, this was accompanied by the realisation of early benefits.

CS4.6 Resource commitment

A five-day MRPII class was held during February 1989 for the management team and the commissioning team was formed. Figure CS4.3 shows the resource allocated to implementation over the two-year period. From Figure CS4.3 it can be seen that a very high level of resource commitment was necessary in order to be effective in driving such a major management change.

Weekly central team meetings were needed, supported by specialist task force activity. Primary task force activities were:

- Inventory record accuracy group.
- BOM team.
- Planning group.

Case Study 4: Kodak Ltd

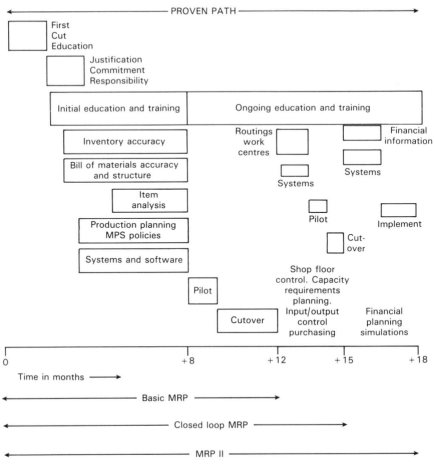

Figure CS4.2 The proven path (*Source*: Wallace (1985))

A 'team pilot' was prepared for July 1989 aimed at resolving major process design and people issues. There was an underestimate in the amount of work involved. This work was associated with the change taking place:

- There were design needs brought about by planning and software.
- Education and training had to be undertaken.
- Capacity had to be managed.

This ever growing list of issues had not been foreseen.

A second two-week full-time whole team pilot meeting was reconvened in September 1989. This required a commitment of twelve to fifteen key people. The

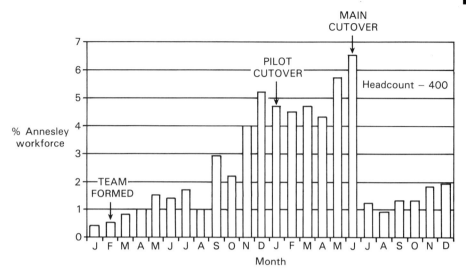

Figure CS4.3 MRPII Activity 1989/90

outcome was an agreed process design and some personnel reorganisations which were more supportive of a better control of the workplace:

- Introduction of a full-time data management officer.
- Improvement in the quality of the resource in production planning.
- Replacement of a reactionary key production manager.
- A decision that there should be a split cutover with a pilot in the roll film unit in January 1990 followed by a cutover for the 135 operation in the following June.

Kanbans were piloted in the interface between the injection moulding and spooling areas.

CS4.7 Roll film pilot

This took place in January 1990 after a dress rehearsal in December. The lessons learnt which would be applied to the main cutover were:

- The need for time fences (see Appendix 2 of this case study on p. 220 for discussion of time fences).
- Leadership from sales operations planning.
- Better operational education and training.
- Involvement of the operations managers.
- Spread the intense data load.
- Involve vendors earlier.

Nevertheless, significant stability, inventory and service gains were seen.

Case Study 4: Kodak Ltd

CS4.8 Preparation for the 135 cutover: January–April 1990

The production and operations managers were now owning the decisions regarding planning, material and information flow within their areas. This was reinforced by two major workshops.

Training activity was at a high level. Much time was spent resolving how material would flow between interfaces. This led to major leadtime reductions.

A major reorganisation of planning occurred at this point, bringing the master scheduling and vendor scheduling functions on site.

A decision was made to phase the cutover by area, e.g. it was planned to convert packing materials in May because a huge data load was involved.

CS4.9 Cutover and instability: June–August 1990

The 135 cutover process took place in June but a period of instability followed due to:

- A previously unknown system 'bug' which saw the failure of the massive 'MRP' run to set co-ordinated plans.
- A last minute response to an Italian request to load the plant immediately to breaking point.
- Insufficient training and understanding by many operations area personnel.

During cutover the old system was decommissioned and for 8–10 weeks there was danger of total system collapse. This was due to incomplete process design around capacity planning and shop scheduling aggravated by poor transaction quality with the new computer system. This led to a recovery programme.

CS4.10 135 cutover recovery phase: September–December 1990

Three task forces were formed to solve:

- A constantly overloaded master schedule due to poor capacity planning.
- An over-complex and mismatched shop scheduling process.
- Endemic transaction errors as people learned how to use the new computer system.

In tackling these tasks the following processes were introduced:

- Line scheduling/135 generic kanbans.
- Rough cut capacity planning.
- Barcoded ticketwork to minimise keyboard errors.

MRPII benefits

```
                SHOP-FLOOR CONTROL SIMPLIFICATION
    Pilot              Main
   cutover            cutover
     |                  |
     |                  |            First form      Flat bill
     |                  | End cap    cassette
     |                  |  kanban    kanban
   Kanban               \    \         \
   pilot                 \    \         \
  (plastics) ─────────────\────\─────────\──────────── Generic
                           \    \         \            kanbans
                            Lid label    Line
                            kanban       scheduling
                  'Brand name' kanbans   Hybrid shop
                                         order kanban
```

Figure CS4.4 135 and roll film flow

CS4.11   Control and class B: January–June 1991

The shop-floor control simplification is shown in Figure CS4.4. In association with line scheduling first form cassettes were removed from the BOM as a level. An immediate upturn in MPS item level service was achieved and by Easter the twelve week average moved from 50 per cent to 80 per cent.

Keeping the line schedule order sequence throughout the flow's generic kanbans proved to be the next hurdle to be crossed.

Refinement of the rough cut capacity planning procedures by production operations managers gave a significant boost to service. The introduction of a production planning meeting as a pre-sales and operations planning meeting was started. This closed the loop with master production scheduling and introduced an effective forum for the communication and control of new product introductions and the disposition of any excess and obsolete materials.

CS4.12   Class A completion: July–October 1991

135 line scheduling and kanbans were paying dividends. Integration of MRPII with financial reporting and forecasting systems was a major thrust during this period. The gains in simpler systems, more accurate data and better production/finance teamwork were greater than anticipated.

The drive to improve vendor relationships and control was particularly effective. The task of publishing MRPII policies and procedures followed.

CS4.13   MRPII benefits

Figures CS4.5 and CS4.6 show the benefits which have been obtained by the introduction of MRPII.

Case Study 4: Kodak Ltd

> Benefits since the main 135 cutover – June 1990 to May 1993
> 1 Manufacturing leadtimes reduced by 70%
>   • Cassettes/spooling/packing = 3 days
>   • Slit and perforate   = 3 days
>   • Kanbans throughout
> 2 Supply chain cycle time
>   • Direct deliveries to regional distribution centres
>   • Direct deliveries to customers
> 3 Inventory reduction by 40%
>   • Raw material and work in progress inventory reduced by £12 million
>   Storage space reduction by 80%
>   • All external storage eliminated
> 4 Item level performance improvement
>   Plus 95% – Class A
> 5 Labour productivity increase by 40% (1989–1993)

Figure CS4.5 MRPII benefits

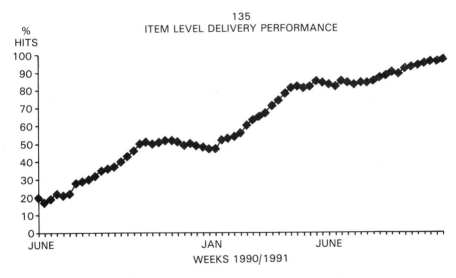

Figure CS4.6 Improvement in item level delivery performance

■ Appendix 1   Class A MRPII user*

Class A uses MRP in closed loop mode. In place and being used are: requirements planning, capacity planning and control, shop-floor dispatching and vendor scheduling systems. Management use the system to run the business by:

- Participating in production planning.
- Signing off production plans.

Source: Wallace (1985).

Appendix 1 C

- Constantly monitoring performance on inventory record accuracy, bill of material accuracy, routing accuracy, attainment of master schedule, attainment of the capacity plans, etc.

The MRP system provides the company plan. All functions use the formal system. There are not manual systems such as shortage lists overriding the schedules. Class A users will rate 90 per cent or more on the following checklist.

1. Technical.

   - Time periods for master production scheduling and MRP are weeks or smaller.
   - Master production scheduling and MRP run weekly or more frequently.
   - System includes firm planned order and pegging capability.
   - The master production schedule is visibly managed not automatic.
   - System includes capacity requirements planning.
   - System includes daily dispatch list.
   - System includes input/output control.

2. Date integrity.

   - Inventory record accuracy 95 per cent or better.
   - Bill of materials accuracy 98 per cent or better.
   - Routing accuracy 95 per cent or better.

3. Education.

   - Initial education of at least 80 per cent of all employees.
   - An ongoing education program.

4. Use of the system.

   - The shortage list has been eliminated.
   - Vendor delivery performance is 95 per cent or better.
   - Vendor scheduling is done out beyond the quoted leadtime.
   - Shop delivery performance is 95 per cent or better.
   - Master schedule performance is 95 per cent or better.

There are regular (at least monthly) production planning meetings with the general manager of staff including production and inventory control, engineering, marketing and finance.

- There is a written master scheduling policy which is adhered to.
- The system is used for scheduling as well as ordering.
- MRP is well understood by key people in manufacturing, marketing, engineering, finance and top management.
- Management really uses MRP to manage.
- Engineering changes are effectively implemented.
- Simultaneous improvement has been achieved in at least two of the following three areas: inventory, productivity, customer service.
- The operating system is used for financial planning.

Case Study 4: Kodak Ltd

■ Appendix 2   Time fences in MRPII*

CS4A2.1   Introduction

All departments must recognise that changes in the plan are time dependent; that is, the closer in the change, the more costly or impossible it may become to make changes in the plan. For every product there are 'time fences' – guidelines that demarcate when changes are feasible. The fences reflect the 'reality of the business'. Figure CS4A2.1 shows the two time fences and three time zones.

- In the time zone outside fence B changes can most easily be made.
- In the time zone between fences B and A materials have been ordered, capacity has been established. Changes in output rates will be difficult to make. Changes in priority caused by change in customer demand and forecasts becoming inaccurate are easier to implement.
- In the time zone inside fence A, even small changes become expensive. It may be impossible to effect changes in spite of extraordinary efforts and costs.

Guidelines have to be established by looking at constraints in terms of capacity and material. Once it can be demonstrated that by sticking to time fences product can be shipped economically, then manufacturing will be expected to work with less time.

Occasionally it will be necessary to make changes within the time fence. If exceptions become the rule the cost of doing business and satisfying customer demands will rise sharply.

Time fences can resolve the conflict between manufacturing wanting long fixed zones and marketing wanting very short response time. This conflict has to be resolved at senior management level.

	A		B	
Emergency time zone		Trade time zone		Plan time zone
Emergency changes		Trading area		Adding and changing
Stabilise		Capacity firm material ordered		Future planning

TODAY ──────────────── TIME ────────────────
──────────────── PLANNING HORIZON ────────────────

Figure CS4A2.1   Time fences (*Source*: Herbert, K. (1992) *Fencing with Time*, Kodak Ltd)

* *Source*: Herbert (1992).

220

Appendix 2

The world class enterprise continuously seeks to find the best balance between:

- High customer service.
- Low cost of manufacture.
- Minimum cycle time.
- Minimum stockholding.

This must be done without sacrificing one for another. The processes for creating and formalising the best balance are:

- Sales and operations planning (S&OP).
- Master production scheduling (MPS).

This must be in association with time fences.

CS4A2.2 Forecast accuracy

If the forecast is 100 per cent accurate:

- Manufacturing would start its purchasing and manufacturing processes to the forecast.
- Shipment of customer order would be JIT.
- No inventory would be held.

Thus the better the forecast, the higher the earnings and customer service.

If the forecast is not 100 per cent accurate:

(a) If the customer is prepared to wait 'x' days and if 'x' is greater than cumulative leadtime (CLT), then make-to-order is possible as above. If 'x' is less than CLT, manufacturing make to forecast, plus extra product to allow for forecast inaccuracy, to the point in the CLT where the time remaining in the supply chain = 'x'. These times can be calculated, and lead to finish-to-order service which is calculated through 'two-level' MPS techniques.

(b) If the customer wants product off the shelf, manufacturing make to stock. Inventory will be held at multiple levels in the supply chain.

As stock is held to combat forecast inaccuracy cost rises. Forecast accuracy can be improved:

- Short-term forecasts are more accurate than long term forecasts.
- Aggregate forecasts are more accurate than item forecasts.

Often end item forecasting is hopeless, because customer demand at catalogue number level is so variable. By moving down the BOM to a semi-finished or component level, items can be identified which will go into many end items. This gives greater stability of demand and allows adequate forecasting accuracy.

Manufacturing costs are minimised when the best stability of supply and demand is achieved. Stability is governed by:

- Forecast accuracy.
- Supply reliability.

Case Study 4: Kodak Ltd

With two-level MPS item forecasts are only needed over short horizons – eight to twelve weeks. These will be used to drive out the last step purchasing requirements. Over longer-term time horizons, (eighteen to thirty months) grade (i.e. aggregate) forecasts can be used.

CS4A2.3 Supply leadtimes

The continuous improvement thrust of manufacturing is to shorten leadtimes. Kodak identified that 1 per cent of leadtime was value adding but 99 per cent was product waiting value-adding activity. Analysis showed that the MRPII initiatives taken could change the ratio of 1:99 to 1:30 at virtually no cost.

To reduce the ratio further would need JIT initiatives:

- Which would reduce waste and expose problems.
- TQC would then be used to effect changes to remove the problems.
- The MRPII environment can be used to plan and control under the new conditions.
- To look for the impact of these improvements on supply chain logistics for further improvements.

Reducing leadtime in manufacture leads to:

- Shorter time zones which improve forecast accuracy. This gives better planning data to manufacturing and reduces the time over which demand must be managed.
- Less safety stock will be required as manufacture moves closer to make-to-customer-order.
- Increased manufacturing capacity becomes available which can be used to meet increased business that sales have the opportunity to gain from shorter leadtime.

CS4A2.4 Setting safety and strategic inventory

Native–CLT (N–CLT), i.e. initial leadtime can be changed into Effective-CLT (E–CLT) through the use of inventory.

- Safety stock is calculated excess inventory to cover the likelihood of actual demand exceeding forecast demand.
- Strategic stock is additional discretionary inventory to cover abnormal (unforecastable demand); or for manufacturing, to cover abnormal supply (supply failure).

It is vital when setting safety and strategic (S&S) stock to consider the whole supply chain. For example, if S&S stock is held in finished goods to cover customer demand variability, then no other level in the supply chain needs inventory to cover that cause.

Inventory drivers, the factors which cause the need for inventory, can be identified. Once they have been identified it is possible to calculate S&S stock. The cost of this stock is an area for continuous improvement.

Appendix 2

CS4A2.5 Managing demand over CLT

Managing customer demand at order entry is fundamental if customers are to be given a delivery performance which can be satisfied. Orders based on forecast become real at order entry. However, the earlier action based on forecast was necessary but managers must remember:

- Forecasts are essential to gear up supply of resources ahead of actual demand so that it can be satisfied.
- Forecasts are inaccurate. Contingency plans can cope with some, but not all, variability.
- Customers never order to forecast.
- Supply capability is finite. Infinite capability is not affordable.
- Supply occasionally fails to deliver.
- S&S stock is finite, because it reduces earnings.

CS4A2.5.1 Order entry

The customer database must be in a form where an order can be identified as:

- Expected in which case material is in the supply chain to meet the demand at the standard order promise service for the item.
- Unexpected, i.e. not in the forecast. In this case demand must be managed as detailed in the following section.

CS4A2.5.2 Managing abnormal customer demand

- Inventory. Policies and procedures need to be established to help order entry determine when, and especially when not, to promise from such stock.
- Leadtime. Part of the order may be supplied from S&S stock. The balance would then be promised to leadtime.
- Capacity. Safety stock and safety capacity (materials, machines and people) can be available to provide a 'fast-track service'. This can offer a shorter promise time than leadtime promising, but this has to be finite because of the cost implications.
- Saying no. Tell the customer the truth if there is not the capability to respond.

One hundred per cent customer service against promise is the customer's minimum expectation and should be a minimum supply standard.

CS4A2.6 Cumulative leadtime, manufacturing leadtime and time zones

CLT and MLT define the capability of manufacturing to supply product in response to a demand for that product. Three time zones were defined in Figure CS4A2.1.

1. Emergency time zone. Within this zone specific product is in process. If

Case Study 4: Kodak Ltd

it is not what is wanted it is too late to do anything about it, without very costly disruption to manufacturing which may be ineffective.
2. Trade time zone. The planner can make most impact on the supply chain effectiveness in meeting demand, by skilfully flexing its capability.
3. Plan time zone. No materials have been ordered. No capacity has been loaded. Manufacturing has its maximum flexibility, constrained only by S&OP production plan implications.

These time zones and their time fences are calculated, based on the reality of actual leadtimes.

- CLT is the time the supply chain needs, through all levels of purchasing and manufacturing, to ship product to meet demand, capacity permitting.
- MLT is the time needed, at a single level, to plan manufacturing, pick, and convert components into parent.

Long leadtimes give problems of demand predictability which destroys the stability needed by manufacturing. The tension between supply and demand must be balanced in the most effective way, whilst striving for 100 per cent customer service, based on 100 per cent delivery to promise. For one major product N-CLT = 154 days. Thus resourcing has to commence five months before delivery. Since there is no way of getting forecast accuracy that far ahead other actions have to be taken.

Figure CS4A2.2 shows the cumulative leadtime for a product which has five process stages. It can be seen that 56 of the 76 days are taken up by purchasing activities. If safety stocks were held to cover 50 days of the purchase leadtime E-CLT would be 26 days; a more reasonable time period for forecast accuracy. However, to do this for all long leadtime products would be prohibitively expensive.

CS4A2.6.1 Vendor scheduling

The way to achieve a 'win-win' result is vendor scheduling. This requires conditions to be created in which the vendor is prepared to share the days of purchased

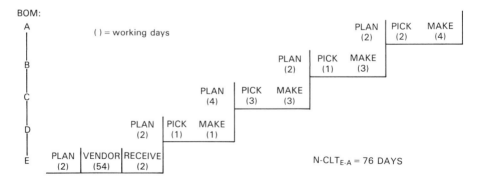

Figure CS4A2.2 Cumulative leadtime through a five stage process (*Source*: Herbert, K. (1992) *Fencing with Time*, Kodak Ltd)

leadtime. In the case considered it may be possible to establish a call-off time of 20 days instead of 56. The key behaviours to make vendor scheduling work are that Kodak:

- Promises never to let demand variability cascade onto the vendor over the agreed call-off time.
- Promises to provide visibility of Kodak demand on the vendor over the vendor's true leadtime – the 'visibility time zone'.
- Promises to constrain the changes in demand on the vendor within agreed tolerances over the 'visibility' time zone.
- Will review with the vendor each other's performance for improvement.

CS4A2.7 Alignment

There is a relationship between time zones at item level and time zones at strategic level.

CS4A2.7.1 The strategic quantification process (SQ)

SQ looks at the direction of Kodak business over the next five years and has time zones.

- The plan zone extends beyond the SQ horizon, i.e. beyond year five. In this zone strategic managers are looking for opportunities.
- The trading zone allows adjustment to be made to the direction in which the company is going. Freedom to alter plans reduces. It is easier to alter plans for year five than year two.
- The emergency zone is year one which is managed by S&OP and is the annual operating plan (AOP).

CS4A2.7.2 The annual operating plan

This is year one of SQ expanded in detail as the AOP and managed by S&OP.

CS4A2.7.3 Sales and operations plan

S&OP must be aligned with SQ. S&OP monitors and guides the business, i.e. sales, operations and financial implications are managed monthly. S&OP has its time zones which align with SQ as follows:

- The S&OP plan zone. This extends five years, i.e. to the end of the SQ trade zone. In the first two years numbers will be at a monthly level, after at a quarterly level. S&OP numbers when summed match SQ numbers because S&OP uses smaller time buckets than SQ, and uses a less aggregated detail, it can more easily reflect seasonal business patterns.
- The S&OP trade zone. This extends from the emergency time fence out as far as the leadtime needed to make major changes to the availability of major resources. The extent will be the length of time to add a new

Case Study 4: Kodak Ltd

production facility; or a new crew; or to line up a vendor for a major increase in output. Moving nearer to the emergency time fence reduces freedom of action.
- The S&OP emergency zone. This zone is governed by the E-CLT of the supply chain and is actually at MPS level. As far as S&OP is concerned, this really is a fixed zone.

CS4A2.7.4 Master production schedule

This is the execution side of the business managed and controlled at item level. There will be the three time zones, aligned with S&OP which was the source of the MPS. Its time buckets are daily, weekly and monthly. Before MPS is ratified, it must be submitted to rough cut capacity planning to ensure that the plan which was OK at family level is still do-able at the item mix, based on the latest demand numbers.

- The plan zone. Extends as far as the end of the S&OP trade zone. At this point in time it is so far out that it makes more sense to manage the business at the aggregate S&OP level than at the item MPS level.
- The trade zone. Extends as far as the E-CLT, and stabilises the supply chain. Trading takes place to help support demand management. It recognises the reality of the materials and processes used to make MPS happen, and the need to serve customers.
- The emergency time zone. Between today and the end of the trading zone is a point in time when the MPS must be finalised, and orders released to the shop floor. This time fence is the boundary of the emergency time zone.

CS4A2.7.5 Fixed assembly schedule (FAS)

The FAS is a sub-set of the MPS and is used to manage customer service. The FAS is the early part of MPS configured as finished goods items. The two time zones are:

1. The configured zone representing released orders to the shop floor.
2. The visibility zone to provide extra time for vendor scheduling.

CS4A2.7.6 The rest of the supply chain

The essential management of the supply chain is at MPS level, where individual unit constraints are taken into account over the longer term through resource planning.

CS4A2.8 Conclusion

Figure CS4A2.3 shows how all levels of Kodak are aligned. Moving from the future towards today means that, at the time fences, the management of the business is passed down to the level of detail which gives the best control mechanism.

Appendix 2

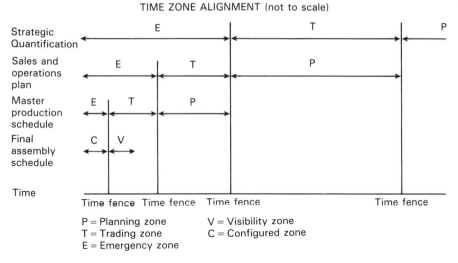

Figure CS4A2.3 Time zone alignment (*Source*: Herbert, K. (1992) *Fencing with Time*, Kodak Ltd)

Time zones and time fences are not arbitrary concepts but can be calculated from the known characteristics of the business. Initial, or native, characteristics of the business can be managed in a way that effectively changes some characteristics to ones which will give a competitive edge.

There are spin-off consequences that call for skilled demand management and 'smart' order promising.

Safety and strategic stock have a specific, calculated part to play, together with vendor scheduling, in underpinning the supply chain.

CASE STUDY 5

Rolls-Royce PLC

CS5.1 Manufacturing strategy in the 1980s

Rolls-Royce is similar to many manufacturing organisations in that it has to integrate new technology with a substantial base of existing equipment. The company operates in a traditional batch environment with a high variety of parts and relatively low individual batch volumes.

Statistics (in 1984):

- Thirty-one engine types.
- Eighty thousand live part numbers.
- Approximately 500 suppliers.
- Approximately 70 per cent of the cost of the engine is spent outside the company.
- Total sales value for 1984 was £1,400 million. Inventory was £841 million.

CS5.1.1 Changes in manufacturing

Recognition of the need to be able to react quickly to design changes and to manufacture these designs efficiently (see Section CS5.3). The way in which the company optimised stand alone machines and attacked metal removal rate was inadequate to remain competitive. Replacing stand alone machines with CNC equivalents was not satisfactory unless thought was given to integration.

A revolution occurred in the approach to production engineering and design and a strategy was set to rationalise the manufacture of components (see Figure CS5.1).

Technical, physical and financial resources were focused on those parts which were critical to the company in terms of engine performance and profitability. Where expertise existed in the supply chain the intention was to utilise that expertise not duplicate it.

The approach to domestic manufacturing strategy recognised the need to concentrate on high technology high performance parts. Figure CS5.2 sets out the domestic strategy.

Traditionally quality had been inspected-in. To achieve right-first-time manufacture an understanding of process capability had to be developed. When process capability was understood there could be a clearer liaison between design and manufacture. This enabled the company to adopt a 'batch of one' philosophy and a target manufacturing time of six weeks. The contribution of a six-week lead-time was vital to enable the company to build to order. As a result a JIT approach was adopted for bought-in items.

Manufacturing strategy in the 1980s

A strategy to rationalise:
- WHAT is made
- WHERE it is made
- HOW it is made
- HOW it is sourced externally
- HOW to control the supply network for lowest cost and highest quality

Figure CS5.1 Manufacturing strategy (*Source*: King (1987))

- High-technology, high-performance, high spares content, components with highest effect on the performance of product and company profitability
- Manufactured in part families
- Manufactured using highest levels of manufacturing technology and automation
- Manufactured in batches of one to maximum of in shop leadtime of six weeks
- Making best use of existing plant and equipment
- Using new working practices in an environment conducive to quality production
- Where quality is engineered into the manufacturing method, sealed after match of process capability and specification

Figure CS5.2 The domestic strategy (*Source*: King (1987))

The purchase strategy which was adopted is shown in Figure CS5.3.

- Sourcing was aimed to be in families of parts so that suppliers could exploit some of Rolls-Royce's advantages of manufacturing.
- Where appropriate, company technology was made available to suppliers for mutual gain.
- Suppliers were to be involved in the risk of new project launch costs. This demanded commitment on both sides, long-term relationships, proven quality and technical excellence.

Use external expertise on those parts where the supply network has gained a particular skill by working with other industries and companies:
- Rationalise number of suppliers
- Source in part families against 'should cost' targets
- Make available all appropriate Rolls-Royce manufacturing technology
  - robotics
  - metal cutting
  - standard tooling
- Make rationalised suppliers part of the Rolls-Royce family
  - quality performance profile
  - risk share of new project launch costs
  - joint buy-off of specifications and process capability match
- Expect 50% reduction on current sourcing costs

Figure CS5.3 The purchase strategy (*Source*: King (1987))

Case Study 5: Rolls-Royce PLC

At this time design was encouraged to talk to manufacturing by a 'How to make' board at director level. Design became increasingly automated and a CAD/CAM approach was being used. Overall these new approaches to manufacturing were aimed at reducing cost, as shown in Figure CS5.4.

CS5.1.2 Integration of manufacture

Group technology is a technique used to group together parts which have similarities in shape size or method of manufacture. Rolls-Royce's grouping into families of parts is based on the way in which parts are manufactured. This led to a study of the way in which turbine blades could be located on fixtures so that changeover between products could be effected rapidly. Linked to tool rationalisation, set-up times were reduced from three hours to fifteen minutes and batches of one became viable. This simplification was needed in order to introduce CIM. The advanced integrated manufacturing system (AIMS) is described in Section CS5.2.

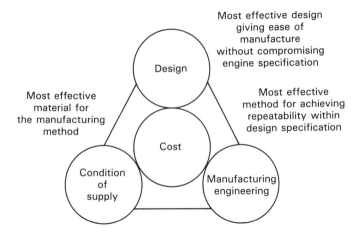

Figure CS5.4 The interactions required to reduce unit cost (*Source*: King (1987))

```
The manufacture of a low pressure turbine exhaust case
                              Before        After
Number of operations            53            20
Manufacturing leadtime         160 hrs       50 hrs
Raw material costs            £2700         £2000
Standard hours content         114            90
Unit cost                     £7300         £5600
Weight saving 5 lbs
Labour saving 50%
Cost of changed method £200,000 which was recovered in 10 months
```

Figure CS5.5 Example of success in new approach to manufacturing (*Source*: King (1987))

Advanced integrated manufacturing system

An example of the gains to be made from simplification and integration are shown in Figure CS5.5.

CS5.2 The advanced integrated manufacturing system (AIMS)

AIMS is a flexible manufacturing system (FMS) for the production of aero engine turbine and compressor discs. It is a self-contained factory within a factory comprising an integrated grouping of versatile machines and process cells that can manufacture a wide range of different disc components under the direction of a computerised central control system.

AIMS comprises 27 cells. Two computer systems in tandem instruct and control the movement of parts between the cells and the automatic racking bay. Components and fixtures are transported by eight automatically guided vehicles (AGVs).

The AIMS project strategy, see Figure CS5.6, was intended to satisfy elements of the Rolls-Royce domestic manufacturing strategy.

CS5.2.1 The goals of AIMS

The specific targets set for the project were to:

- Cut work in progress by two-thirds.
- Reduce production leadtime from 26 to 6 weeks.
- Increase productivity by over 40 per cent.
- Achieve sufficient savings to cover the project investment costs in its first full year of operations.

---

1. Parts should be organised into family groups
2. Parts should be planned for a common manufacturing method with a standardised sequence of operations
3. Tools should be rationalised and standardised
4. Set-up times should be reduced to provide a 'batch of one' capability
5. Work holders should be standardised
6. In-cycle inspection should be provided
7. All manufacturing processes should be integrated into the work flow and handling system
8. Non-machining operations should be examined with a view to applying AMT where viable
9. Work flow and handling should be automated
10. Quality should be improved by matching engineering specification to process capacity

---

Figure CS5.6 AIMS strategy (*Source*: Butcher *et al.* (1987))

Case Study 5: Rolls-Royce PLC

CS5.2.2 Changes effected before the implementation of AIMS

Methods development

A new method of manufacturing discs was introduced in 1979 involving the design and manufacture of a number of four-axis CNC turning centres which could turn both sides of a disc diaphragm simultaneously. This effectively eliminated additional removal operations with the risk of distortion.

These machines reduced the number of operations from 21 to 5 and the lead-time from 22.5 to 8 weeks. The machine tool population was halved (62 to 31) and the scrap rate was reduced by 40 per cent.

Supply of purchased materials

Prior to the changed methods of manufacture outlined above 60 per cent of material purchased was swarfed during the manufacturing process. The new machining techniques obviated the need for excessive material. This allowed purchase of slimmer forgings resulting in purchase savings of £1m per year at 1985 values.

Refining the method

Attention focused on provision of a batch of one. AIMS was to work with a batch of one which required a successful policy of 'design for manufacture'. Prior to AIMS batches were units of ten dictated by high machine setting times, high movement costs, and practical scheduling and shop control limitations.

The techniques used to reduce setting times are shown in Figure CS5.7.

Integration of machines and process equipment

The integration for AIMS involved rearrangement of the existing equipment into 27 cells. Over 40 machines were deployed in 10 of the cells. The remaining 17 cells contained processes such as etching, descaling, cleaning, X-ray and non-destructive testing.

---

1. More refined group technology involving classification of specific features on similar discs
2. Fixture standardisation and control
3. Use of standardised 'rings and sleeves' and adaptors which effectively palletise the discs
4. Standardisation of cutting tools which were reduced from 2000 to 100 standard tool sets
5. Provision of more powerful software packages for the CNC machines (and hardware extensions in the processors to suit)
6. In-cycle inspection
7. Introduction of quick change 'cassette' tooling

Figure CS5.7 Techniques used to reduce setting times (*Source*: Butcher *et al.* (1987))

Advanced integrated manufacturing system

The greatest challenge during this phase was maintaining production amid the change. A greenfield site factory was considered but the decision was taken to locate the AIMS process in the existing factory.

Automation of work flow and handling

This phase included a fully automatic racking area for components and fixtures, with 2 auto-stacker cranes, 4 conveyors, 1 profile gauge and 8 AGVs. Software for the transport control system (TCS) was linked to the AGVs through a network of 21 local traffic microcomputers. Receipt and dispatch were via 45 docking stations.

Control of AIMS

The control of this integrated system was provided by a central control computer system (CCS) which is aware of:

- What is happening.
- What should be happening.
- How to make things happen.

The CCS links in to the company mainframe, which carries the mechanised work booking system, to TCS and to the cells. Its functions are shown in Figure CS5.8.

CS5.2.3 Conclusions

- Total cost was £8.2m.
- The decision not to build a new factory was correct. Less than one-tenth of the potential spend was needed.
- Planning in great detail was amply rewarded. Forty man years of planning went into the final phase of AIMS.
- The size and quality of software for an 'intelligent' FMS should never be underestimated.

---

1. Recording progress of batches through their manufacturing sequence via the link to the mechanised work booking system
2. Maintaining shop status information on machines, work-in-progress and docking stations
3. Managing and scheduling the transport control system and instructing required movements
4. Downloading priorities and operation sequences from the company mainframe
5. Planning and instructing the loading of machines and instructing tool kits

Figure CS5.8 The functions of the central control system (CCS) (*Source*: Butcher *et al.* (1987))

Case Study 5: Rolls-Royce PLC

CS5.3 The development of simultaneous engineering (SE) at Rolls-Royce

Simultaneous engineering attempts to optimise the design of the product and manufacturing process to achieve reduced leadtimes and improved quality and cost by the integration of design and manufacturing activities and by maximising parallelism in working practices.

CS5.3.1 Scope of simultaneous engineering

Effective SE involves all areas of research, product and process definition, and product proving. Figure CS5.9 shows the integrated design and manufacture process at Rolls-Royce.

Both design and manufacturing are involved in the development concepts at the beginning of an engine project. While design works on conceptual and configuration issues, manufacturing provides new capabilities for the technologies which will be required for the new engine.

As the engine design develops, component design will produce specifications for modules and components within the engine, This will involve liaison with manufacturing on materials and tooling requirements.

As the engine configuration evolves, detail definitions are produced for the components and their manufacturing processes. There is a formal acceptance of

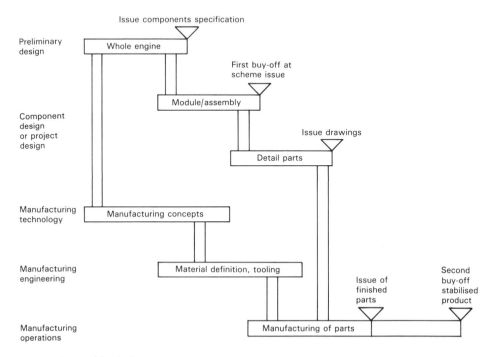

Figure CS5.9 Simultaneous engineering – integrated design and manufacture phase (*Source*: Broughton (1990))

Simultaneous engineering (SE) at Rolls-Royce

> 1 Design verification
>   - Demonstration of predicted behaviour
>   - Instrumented rig and engine testing
>   - Manufacturing process verified
> 2 Certification
>   - Demonstration of airworthiness
>   - Failure tests, type approval '150 hr' test, o/speed and o/temp tests etc.
> 3 Maturity
>   - Cyclic endurance testing
>   - Representative of service operation

Figure CS5.10 Product validation (*Source*: adapted from Broughton (1990))

the engineering specification by manufacturing, indicating that they agree that the design can be made and that there is a match of process capability and the engineering specifications.

Integrated design and manufacture for an aero engine are a vast operation. Control of leadtimes, quality and cost can be achieved by the use of SE techniques. Product validation is an essential part of bringing an engine to market. Figure CS5.10 outlines this process. Any action resulting from product validation will be executed using SE.

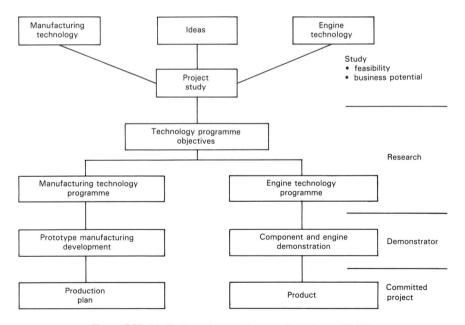

Figure CS5.11 Project phases (*Source*: Broughton (1990))

Case Study 5: Rolls-Royce PLC

CS5.3.2 The management of simultaneous engineering

The management in Rolls-Royce is by three groups detailed below.

Technology acquisition group

Areas for improvement in engine technology, and the resultant improvements in manufacturing technology have to be identified. This activity is carried out by a high-level multi-functional group which monitors market requirements for future engines.

This group will direct concurrent engine demonstrators, component, material and process development. The scope is shown in Figure CS5.11.

Engine project management group

A multi-disciplined team of managers is set up by, and collocated with, the chief engineer. This enables fast communications between these key personnel and an overall awareness of the problems faced by the team. This structure is shown in Figure CS5.12.

The design/make process, the engineering natural group

Figure CS5.13 shows the variation in group composition at the various stages of the product life cycle and underlines the transience of managing SE. To ensure maximum parallelism of activity between groups, technical and programme meetings are held at regular intervals. These groups are not collocated and initially are led by design. The aim is to identify leadtime items, and critical data flows between functions as well as ensuring that project milestones are achieved.

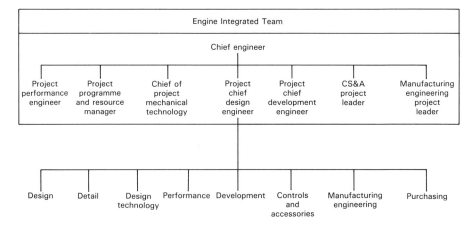

Figure CS5.12 Integrated team structure (*Source*: Broughton (1990))

Simultaneous engineering (SE) at Rolls-Royce

	Advanced design	Detailed design	Product proving	Service fleet
Chief engineer	*	*	*	*
Mech, tech	*	*	*	*
Performance	*	*	*	*
Project design		*	*	*
Aerothermal	*	*	*	
Development		*	*	*
CS&A		*	*	*
Component design	*	*		
Manufacturing eng.	*	*		
Advanced design	*			
Head of project	*			
Marketing	*			
Experimental			*	

Time →

Figure CS5.13 Group variation with phase (*Source*: Broughton (1990))

Technical and programme problems are quickly highlighted, enabling a fast response by re-allocation of resource as required.

Classes of design task Analysis of the design process has identified three classes of design task:

1. Runners. Defined as designs where existing experience, methods, databases and rules can be applied. In these cases the detailed design definition can be performed with minor input from specialist functions.
2. Repeaters. Defined as designs where existing experience and methods can be applied but detailed design definition requires significant input from a wide variety of specialist functions.
3. Strangers. Defined as a complex, high cost design containing a high degree of novelty. Detailed design definition requires significant and often novel input from specialist departments. Time scales become difficult to plan because of inherent uncertainties and a high degree of management attention is needed.

The classification allows decisions to be made on the formation of transient groups to handle the design of more complex components. How far this can, and should, be taken depends on the availability of trained, specialist resource.

At the end of the product definition phase which considers what should be made and how, there is a formal acceptance by the manufacturing area of the manufacturability and cost of the design which has been specified.

CS5.3.3 The management of change

Figure CS5.14 presents the changes to be managed by Rolls-Royce in successfully implementing SE.

Case Study 5: Rolls-Royce PLC

- New methods meant that engineers had to change well-established working practices
- Integrated project teams involved the collocation of managers
- Engineers and technololgists now have a sense of identity with projects
- For particularly difficult design and manufacturing tasks engineering and technology are collocated
- The advantages and benefits of collocation must be quickly recognised by the project teams, if they are to react favourably to the breakdown of traditional departmental demarcations
- Engineers are expected to use new computer modelling systems
- SE needs excellent leadership from managers and the workforce must be willing to change

Figure CS5.14 The changes facing Rolls-Royce in adopting simultaneous engineering (*Source*: adapted from Broughton (1990))

CS5.3.4 Computer aided engineering and manufacture

Rolls-Royce has formed a corporately based computer aided engineering and manufacturing strategy. This has three main areas of activity:

1. Product and process definition. This is concerned with generating, analysing and validating the shape, properties and manufacturing processes associated with a product. For many of the components within an engine the integration of design and manufacturing modelling systems proves very difficult. The large variety of tasks presents a significant problem in finding a satisfactory computing database.
2. Product definition control. This is concerned with storing, protecting, communicating and monitoring the product and process definition data as it flows through the definition processes. The volumes of data transferred are large, the time taken is non-value added.
3. Product and process definition information support. This aims to provide the engineer with the relevant information to perform the tasks. Rolls-Royce totally integrated engineering and manufacturing databases are called 'key systems'. These have been developed for disks and shafts, compressor blades and installations with others to follow. The disks and shafts key system enables the manufacturing engineers to plan the manufacturing sequence from the design model of a disk or shaft. The same model will back-track from the design intent through turning, broaching, machining and finally to the forging requirement. The system has not only reduced leadtime but also resource requirement.

Areas not connected through common database systems are covered by a drawing scanner system which enables drawings to be electronically recorded and sent anywhere in the company. Modifications can be made to the drawing on a work station or the drawing may be plotted out, altered and then re-scanned. As a result design leadtimes are reduced and, manufacturing engineers have an easy access to make modifications to proposed designs as they evolve. This system is being further developed as progress in information technology changes the state of the art.

CS5.3.5 Examples of simultaneous engineering

The phase 5 combustor

During take-off this will be expected to accept air from the high pressure compressor at 570° C and 35 times atmospheric pressure. Fuel will be pumped in and atomised before burning at 2,500° C. Typically heat release rates are 100 MW in a volume of 85 litres.

The combustion engineer must take into account a large number of requirements that are placed on the combustor (see Figure CS5.15). Paramount are those requirements which impact on airworthiness.

As new combustion cans are designed fairly infrequently, this new can has been treated as a 'stranger' and a large project team of specialists and detailers were collocated with the designers. A common database between designers, detailers and specialists was used to enable rapid design iterations in this very short leadtime project.

Manufacture of this product is achieved primarily by the supplier network and, hence, this project has involved suppliers at a very early stage in the design phase.

By implementing SE in this project conventional leadtime has been reduced by 30 per cent. Costs and weight have been better controlled. Reduction in conventional leadtime has been achieved.

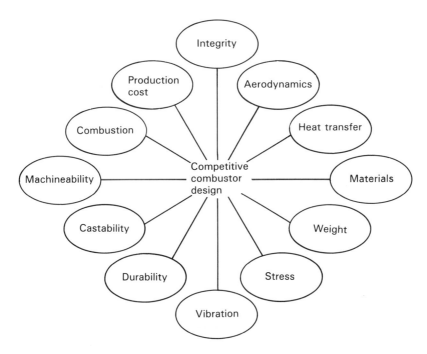

Figure CS5.15 The requirements for the phase 5 combustor (*Source*: adapted from Broughton (1990))

Case Study 5. Rolls-Royce PLC

The wide chord fan blade

The RB211 fan blade is a precision fabrication between 25 and 40 inches long made in titanium alloy. At aircraft take-off a fan will weigh about 75 tonnes and soak up over 1400 horse power.

The first generation blades were a fabricated assembly consisting of a honeycomb centre between two titanium sheets diffusion welded at the edges. The second generation blade for the Trent engine is a diffusion bonded/super plastically formed construction and was treated as a 'stranger'.

Since the design centre is located a hundred miles from the manufacturing centre high levels of integration between product and process definition were difficult to achieve. To solve this problem, the nominated multi-disciplinary team met weekly at the design or manufacturing centre dependent on the relevance of location to the discussion required.

By employing SE, the second generation fan has been designed and manufactured in the same time as a fan based on the first generation configuration. Projected cost has been maintained and the weight of the fan is approximately 25 per cent less than a scaled first generation fan.

CS5.3.6 Summary

- By using SE Rolls-Royce is gaining competitive advantage in its integrated engineering activities. The advantages are in terms of cost, leadtime and quality (see Figure CS5.16).
- Design and manufacturing information is held which can be used for conceptual engine work at preliminary design. Conceptual design is followed by detail design controlled by the design/make procedure. Before manufacturing commences there is a formal acceptance procedure.
- Highly structured product validation enables feedback of engine test data into design for verification of design assumptions.
- The engineering activity for an engine project is centred about the chief engineer who will manage an integrated, collocated team of managers, responsible for the optimal resourcing of the engineers allocated to them.
- SE represents a change to the roles of management and engineers who need training in new working methods. Without this, SE will fail.
- SE requires high levels of communication between participating groups. This leads to the formation of integrated teams and, if high levels of interaction are required, collocation.

CS5.3.7 Developments in the 1990s

In developing SE, Rolls-Royce has become increasingly aware that the cost of manufacture is very dependent on the specifications and, therefore, of the importance of engineering working effectively with both domestic and external manufacturing suppliers in achieving a competitive product specification.

Systems engineering techniques have been used to gain an understanding of the subprocesses involved, with the aim of creating a structure which provides better accountability and optimisation. This has shown that the two major

Simultaneous engineering (SE) at Rolls-Royce

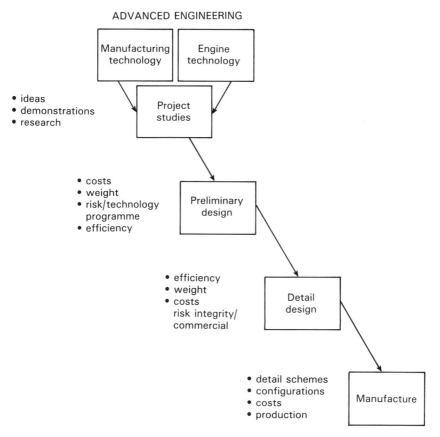

Figure CS5.16 Simultaneous engineering in aero gas turbine design and manufacture (*Source*: adapted from Broughton (1990))

subprocesses requiring optimisation are:

- Acquisition of capability which provides for the future requirements of engineering in terms of the technology, test facilities and people that will enable the right products to be produced and verified on time.
- Product definition and verification. The early development shown in Figures CS5.9, CS5.10 and CS5.16 has evolved into a sequence of six subprocesses:

define whole engine
define modules
define components
verify components
verify modules
verify whole engine

Definition and verification of product and process integrate manufacturing. Product definition has to be in line with the process capability of the

241

Case Study 5: Rolls-Royce PLC

manufacturing process to be used, irrespective of whether this takes place in domestic or external facilities.

Organisation

Collocated function based teams (Figures CS5.12 and CS5.13) have been replaced by integrated engineering teams.

Propulsion systems engineering initiates a new product concept. Research objectives for new design and manufacturing technologies are identified. The task is broken down into subsystems (e.g. fans, compressors, combustors, turbines) and the requirements of each module are defined to include weight, cost, performance, etc. The process is repeated again as the modules are broken down into individual components.

Manufacturing is involved throughout and the process is facilitated by the computer aided engineering environment.

Computer integrated engineering/advanced technology

The product definition process is simplified and integrated by all engineers working on one model, where research data is translated into:

- A design code.
- A control program.
- A performance model (for verification in an experimental engine prior to certification).

Computer systems which interface at all levels of the SE process are described as 'key systems'.

Technology acquisition and materials development

The process described in Figure CS5.16 have evolved. SE is based on the need to develop technology into models and data bases and to develop materials which will meet the demands of new designs.

CS5.4 Conclusion

The case study presents stages in the evolution of SE. SE is currently regarded as a mechanism for optimising the use of innovative technology in the manufacture of aero gas turbines. To integrate manufacturing requires extensive planning and the development of relationships between a team of different people under the control of several different organisations in three continents.

Time scales are reduced by considering the minimum information needed to start each activity, followed by data to keep the activity going. Planning and control are achieved through the formation of an integrated team of technologists, engineers, detail draughtsmen, manufacturing engineers and, critically, the suppliers of the parts. Communication is facilitated by computing facilities and other data transfer devices.

Acknowledgements

Acknowledgements

Acknowledgement is made to the following sources for sections of this chapter:

CS5.1 D. King (1987) Approaches to manufacturing, in *The Management of Manufacturing* (eds R.P. Toone and D. Jackson), IFS Publications Ltd.

CS5.2 M.C. Butcher *et al*. (1987) An advanced integrated manufacturing system for turbine and compressor discs, reprinted by permission of the Council of the Institution of Mechanical Engineers from *The Proceedings of the Institution of Mechanical Engineers*, Vol. 201, No. B3.

CS5.3 T. Broughton (1990) Simultaneous engineering in aero gas turbine design and manufacture, *Proceedings of the 1st International Conference, Simultaneous Engineering*, Status Meetings Ltd.

CS5.3.7 T. Broughton, Principal Fellow, Warwick Manufacturing Group, University of Warwick.

CASE STUDY 6
Stanley Tools

CS6.1 Introduction

Stanley Tools makes a wide range of traditional woodworking tools. The range comprises over 4,000 different products. The company competes on the strength of its brand name which is taken to be synonymous with quality. A recent survey showed a brand awareness of 93 per cent. This image is supported by constant product development which is driven by a team approach making use of both product and process innovation.

Competition from Third World countries is intense both in own brand products and in thinly disguised copies. Until recently Stanley has been opposed to own label manufacture. With the increasing importance and power of DIY chains, this attitude is changing. Retail chains such as Marks & Spencer or the John Lewis Partnership have their own brand labels, St Michael and Jonelle and are responsible for the product and quality specification. In contrast DIY chains, generally, depend on the manufacturer's specification. Thus, Stanley is having to develop product specifications for DIY chains, which will be mutually beneficial, and also competitive with Third World products. Within this scenario the Stanley branded name will continue to compete as the value for money quality product.

This case study presents four examples of the way in which Stanley is purposefully managed to be successful in achieving competitive advantage.

1. Product development.
2. Manufacturing planning and control systems.
3. Process development.
4. Organisation in manufacturing.

CS6.2 Product development – the Magnum screwdriver

An excellent example of Stanley's strategy of developing branded products is the Magnum range of screwdrivers.

The development of this product commenced through a multi-function management team, established to develop screwdriver business. Multi-function management teams are the normal means of initiating such activity.

The proposal was to introduce a range of screwdrivers into the domestic and mainland European markets. The latter is highly lucrative but intensely competitive. The range of screwdrivers was to be designed to satisfy the needs of DIY and professional users in both the wood and engineering aligned trades.

CS6.2.1 Design considerations and specification of the Magnum screwdriver

The product discrimination between the two professional trades was a prime design consideration.

The wood trades' demands were for a comfortable handle capable of transmitting both a substantial end-load and torque associated with the insertion and removal of stubborn wood screws.

The engineering trades' preference was for a handle form suited to the manipulation and, maybe, precise adjustment of machine cut screws into pre-tapped holes prior to the final tightening torque. It was important, however, that the handle should be capable of transmitting torque values comparable with the traditional extruded-style handle without undue discomfort when used in an oily/greasy environment.

It was a prime requisite of all users that the quality and precision of screwdriver tip geometry should be of a standard that would ensure a good level of performance under a variety of operating conditions.

The handle specification

The handle specification was the innovation that would give Stanley competitive advantage. Handles are based on a common tri-lobal geometry incorporating a rubber element at the peak of each lobe to give soft touch/high torque characteristics. The largest proportion of the handle is moulded in polypropylene with its inherent strength properties. Additional circumferential soft touch areas are included, the rearmost incorporating the 'Stanley' logo on all three faces.

The handles are injection moulded, based on the two colour moulding technology. This technology was pioneered at Stanley with the introduction of the moulded Thrifty screwdriver handle. Development of the technology took place in conjunction with machinery suppliers Klockner Ferromatik Desma Ltd and toolmakers Braun & Keller GmbH.

The advantages gained from using available technologies to the full are to offer a handle form acceptable to all sectors of the European screwdriver market and to give Stanley a unique product that is difficult to copy.

Bar specification

The total range is 111 covering flared, parallel, crosspoint, torx, awl, nut runner and hex drive variations. The Magnum screwdriver bar was originally conceived as being the existing Stanley standard lacquered bar but with the tip black etched to mimic the established European product. (European producers chrome plate and mask the working end to prevent flaking.)

However, because Stanley bars are lacquered, and tip definition is sharp, a further development took place that resulted in a specification giving a lightly shot blast tip. This tip provided an added-value feature of increased screwhead grip.

Case Study 6: Stanley Tools

CS6.2.2 Marketing

Domestic marketing strategy

Since 1987 there had been a decline in the volume of Stanley fixed blade screwdriver sales. This was in both the retail sector where the threat was from the many imported lookalikes, and on the industrial front where price competition was being met.

The Magnum was positioned in the UK towards the industrial sector, i.e. the 'greasy hands' trades but it was expected to satisfy both the wet and dry users. The introduction would allow repricing of existing models.

European marketing strategy

A fundamental technical requirement was to create a product with high torsional strength to enable serious competition in the European market.

The European screwdriver market in 1988 was worth $153 million. Three German and one French company accounted for 28 per cent of this market with non-European imports making up a further 47 per cent. Stanley Tools with sales of $8.4 million was ranked fifth with a market share of 5.5 per cent. The four major competitors are specialist screwdriver manufacturers without a broad line of tools and, for the most part, have limited penetration outside their own domestic market.

With the addition of the Magnum range of screwdrivers Stanley Tools was offering a selection of screwdriver tips as wide as any available in Europe, featuring black tips, ferromilled bars and most significantly of all, the two material injection moulded handle.

Packaging

Packaging is an important feature in the retail marketing of Stanley products. It was proposed to offer two packages:

1. Bulk packaging with six screwdrivers to a box.
2. Carded, where the product is 'stapled' to the card, the handle pointing down and protruding beyond the card to allow customers to feel the product.

CS6.2.3 Capital justification

The project was expected to provide a high ROI if anticipated volumes of sales were achieved. To evaluate the risk involved in the project sensitivity analysis was carried out to identify the effects of two factors:

1. Substitution of an existing range in a major market.
2. Failure to achieve target volumes for the Magnum screwdriver.

The ROI that would result from a best to a worst scenario was calculated. It was decided that the project carried an acceptable degree of risk and the capital expenditure was authorised.

CS6.2.4 Achievement

Since their launch the Magnum series of screwdriver units have increased by 12 per cent with turnover value increasing 44 per cent.

CS6.2.5 Manpower and costs

Magnum handle moulding is undertaken at the Hellarby plant and comprises five Klockner Ferromatik Desma two colour moulding machines.

The Hellaby moulding department comprises twenty-two moulding machines used for making a wide variety of Stanley moulding requirements. These include the Magnum machines. The department works a 24-hour, 3-shift system with each shift comprising one technician, one operator and one setter/operator. The operator and setter/operator are classed as direct workers who operate under incentive conditions (payment by results – PBR). The technician also supervises.

Manufacturing costs are established using standard costing methods with screwdriver handle moulding machines (including non-Magnum) having a separate overhead rate from other moulding machines.

Performance is measured using conventional off-standard reporting.

CS6.3 Manufacturing planning and control systems

CS6.3.1 Scheduling and sales order input

Operations are controlled using a three-month rolling master production schedule (MPS) updated monthly. The primary schedule is computer-generated and is based on historical sales, current forward orders and finished goods stocks. The primary schedule is reviewed line by line to take account of known likely deviations, such as promotions, expected large export orders, etc. The revision is carried out jointly between manufacturing and stock control. The resultant MPS is loaded into the system and forms the basis for materials purchasing and materials requirements planning.

Sales orders, with due delivery date, are entered on receipt and update the order book.

CS6.3.2 Manufacturing control

The MPS is available live on VDUs in the manufacturing locations down to foreman level. This live information is supplemented by weekly printouts showing:

- Analysis of the order book by line. This identifies the current stock/backlog situation together with an analysis of the forward order situation split into two-week segments.
- Backlog report showing products in sales backlog in descending order of value.

The availability of this information to the manufacturing units, together with the live, on-screen information is critical to maintaining good customer service together with controlled finished goods inventories because it allows the

Case Study 6: Stanley Tools

manufacturing department to deviate from the MPS *when it is sensible to do so*.

Sales order and schedule monitoring

Monitoring of the immediate schedule versus current order book is done on a daily/weekly basis by the team responsible for the production of a product group. This is supported by a human relations management policy of product ownership by the team making the product.

The review of the schedule requirement is an ongoing activity using the on-line and printed information available. For instance, a typical week would begin with a review of the weekly backlog report and the analysis of the order book. Items in sales backlog or potential backlog can be quickly identified, and can be noted for closer monitoring of the production schedule and the daily incoming business.

The following are some of the factors that would be used to determine whether the schedule should be changed:

- How long will the product be in backlog or short supply if no changes are made?
- Is the item a fringe or mainline product?
- Will the change to schedule have a serious detrimental effect on other manufacturing priorities?
- Will the item shortage cause serious problems in other areas, e.g.
    - a large export shipment will be held because this product is not available;
    - shortage will cause sales embarrassment with an important customer;
    - letter of credit expiry on export orders?
- Will immediate schedule changes have a serious effect on manufacturing efficiencies?

Enforced changes to the manufacturing plan can be caused by machine breakdown, etc. In this event the availability of on-line information allows management to determine the best course of action. For example, if the immediate demand is urgent should production continue by a more costly off-standard method?

Stanley believes that the factors influencing the day-to-day shop-floor activities require a delicate judgmental balance that is best made at factory level providing that the sort of information detailed above is made available.

CS6.3.3 Summary of manufacturing planning and control at Stanley

In essence the system is used to process relevant information so that experienced personnel can make decisions, This promotes an interactive relationship between the need to maintain a high standard of customer service on an ongoing basis coupled with an efficient factory operation.

CS6.4 Process development

Process development takes place in conjunction with new product development as outlined in the introduction of the Magnum screwdriver. However, process development is carried out for a variety of reasons:

- The manufacturing processes are hazardous and development is needed to improve safety, e.g. the elimination of lead bath hardening.
- Job enrichment.
- Cost reduction.
- Quality improvement.

Two process developments are considered, each involving the application of robots to existing processes. Technical difficulties were encountered in the introduction:

- Achieving planned cycle time.
- Engineering each robot to communicate with peripheral equipment effectively but safely.
- Achieving reliability and consistency of peripheral equipment which, unlike the robot, was a one-off design application.

CS6.4.1 Claw hammer head forging

This operation involves an operator picking up a red hot billet, using tongs. The billet is presented to a series of forging and clipping operations which form a claw hammer head. Forging is carried out on each of two forge cells by a three-man team, with each operator physically forging 40 minutes in every hour as part of a two-man team. The floor-to-floor time is 8.5 seconds.

Net capacity for a 39-hour week using both forges is 21,600 heads.

The operatives work in hot, dirty and noisy conditions. Their work is monotonous. The operations performed are the transfer of material from one operation to another. Stanley considered that this was an operation where labour could be replaced by a robot.

The development of a robot and assessment of its viability

In conjunction with ABB Robotics a robot was designed, which was successful in transferring the billets through the forging and clipping operations. The robot achieved a floor-to-floor time of 31 seconds. This time could be reduced to 23 seconds by further process modifications. There was no possibility of equalling the output from the manual operation which, if one operative was employed, would give a floor-to-floor time of 17 seconds.

By retaining clipping as a manual operation and allowing the robot to feed the head forming operation only, an output of 17,300 heads per week could be achieved. This, however, would have resulted in:

- The retention of operatives in the hostile environment.
- An increase in unit labour cost.

Case Study 6: Stanley Tools

- A reduction in output of 4,300 heads per week.

The result was that the manual method of operation was retained.

CS6.4.2 Robotic hammer head heat treatment

Heat treatment originally involved lead bath hardening. This process presented a health hazard and was replaced by induction hardening and oil quenching. The change in process gave an opportunity to further develop the process by the formation of a robot cell.

The working of the robot cell

The hammer heads are fed in pairs to the robot from a hopper. The robot presents the pair of hammer heads to the induction heating elements, at the same time unloading the two already treated and placing them in the oil bath. The robot unloads the pair of heads from the previous cycle and returns to the start.

This simple description represents the outcome of considerable investment of resource in solving the problems of presentation and timing. The robot cell processes standard, high-volume hammer heads. A manually fed line for non-standard heads run in parallel. One operator is responsible for the robot cell and the manual line.

The operators PBR is based on the total output from the robot and the manual operation. The operator has incentive to ensure the smooth running of the robot operation as well as achieving maximum output from the manual line.

Achievement

The achievement is:

- A modest saving in cost.
- Elimination of a hazardous operation.
- For the operator, replacement of a manual dexterity skill with a technical skill.
- For the volume hammer head production, a very carefully controlled operation giving optimum quality.

CS6.5 Organisation in manufacturing

CS6.5.1 Team working

A strength of the Stanley organisation is the product line management team. Each product line effectively becomes a strategic business unit and the team has many of the characteristics of simultaneous engineering when new products are developed. The management control system reports performance at product line level. The organisation at product line level is supported at company level by both product and process development activities.

Cell manufacture is a norm at Stanley. The company would claim that cell manufacture has been in existence for more than 30 years. The cells are not the U-shaped cells generally advocated. Walking round the plant it is difficult to identify the cells. The cells are groupings based on the operations needed to make a product. Successive operations are controlled by kanbans, i.e. a machine is operated to fill a container. When the container is full production ceases until the container is removed by the next operation. Cells are flexibly manned by operatives who can perform all the operations in a cell.

An important factor in the successful operation of the manufacturing process is the strength of the Stanley culture. Personnel at all levels are aware of what is happening in the organisation and the issues that are involved. At shop-floor level this is brought about by, for example:

- Supervisors' ability to access the sales order/production control system via computer terminals. This allows them to foresee changing priorities brought about by a dynamic market and plan an appropriate reaction.
- The retention of a payment by results (PBR) structure that has been modified to support the achievement of goals.

CS6.5.2 Payment by results scheme

Stanley considers that the retention of a PBR scheme is necessary where work is repetitive, fairly simple and essentially dependent on operator effort.

Job evaluation

Six factors are considered:

1. Working conditions.
2. Physical effort.
3. Responsibility.
4. Skill.
5. Supervision of others.
6. Safety of others.

Each factor has a weighting.

Job evaluation results in four production grades across the factories. These range from the lowest paid grade of General Operator, Assembler or Packer to the highest paid grade of Skilled Machine Operator.

Flexibility exists between grades subject to the operator's ability, having been provided with training. Operators are paid the higher grade for work carried out, if they are transferred.

Work study

All direct operations have been studied and have been given a standard time. Stanley industrial engineers use British Standard (BS) techniques to measure operations. They use the Tel-time System for data capture which gives ease of calculation, computer storage and printout.

Information on standard times is available to the operator, shop steward and foreman. Any discrepancy can be discussed at this level and resolved. If the problem is not resolved there is an appeals procedure. All shop stewards have been instructed in the use of the Tel-time System in particular and BS work study measures in general. All should be aware of, and can measure, what a standard performance level actually is.

CS6.5.3 Incentive schemes

Base rate of pay (datal) is paid for all hours worked and is the starting point for bonus at 75 performance. One third bonus is earned at the standard performance of 100 and the scheme follows a straight line to its cut-off at 120.

Management uses the bonus scheme to promote flexibility and transfer operatives away from cells where production is not needed to cells where increased output is a priority. Waiting time is paid at datal and the only way for operatives to get bonus pay, if production is not required in their cell, is to be transferred to undermanned cells where production is needed.

Incentive schemes have been developed which are not individually based PBR schemes. There are group schemes where work is highly variable, for multi-machine operations, long cycle times and machine-controlled operations. Generally the incentive is to keep machines operating and the incentive element of pay is reduced. Over recent years there has been a move to group schemes based on cell working where operators are paid on the output of the cell.

CS6.5.4 The management's assessment of what tools and techniques to adopt

Stanley is a company that examines its environment and adopts tools and techniques from which it considers it can obtain value for money.

CS6.5.5 Total quality management

BS 5750 is in place. Quality is highly regarded and, at each machine, operatives have displays of the inspection requirements to meet quality standards. But the company has not attempted to develop their acknowledged drive for quality into TQM. Their view is that the cost/benefit does not support a change in their system of achieving quality. Little use is made of SPC. Reliance for quality is based on 'ownership of the product' by the shop floor.

CS6.5.6 Maintenance by operatives and multi-skilling of maintenance

Tactical WIP stock is held to guard against machine breakdown. Manufacturing management has decided that the cost of planned preventive maintenance outweighs the benefit. Operatives have not been involved in maintenance of their machines beyond machine cleaning. Multi-skilling in the maintenance force is being considered for the future but there are reservations about how far multi-skilling can be taken with the sophisticated electronic controls on many machines.

CS6.5.7 Continuous improvement teams

There is a suggestion programme but no continuous improvement teams of operatives. Product line management teams consisting of European and local marketing, the production manager and the development engineer are responsible for this activity. Equipment replacement is also used as an opportunity for enhancement.

CASE STUDY 7

The Tempered Spring Company Limited

CS7.1 Background

The Tempered Spring Company Ltd (TSCo) is a subsidiary of T&N Group and manufactures a comprehensive range of spring products by the manipulation of high alloy steels. This case study is concerned with the division manufacturing valve springs.

CS7.2 Product – engine valve springs

Valve springs are used in four-stroke internal combustion engines. The action of the camshaft opens the valve and at the same time, compresses the valve spring. The stored energy in the spring is then used to close the valve as the spring expands.

CS7.2.1 Future developments

The needs of the market demanded a continuation of existing developments which are shown in Figure CS7.1. In response to these needs there was clearly potential to develop new valve spring materials and/or processes which would give improved load characteristics and smaller spring sizes.

TSCo analysed springs by means of finite element analysis and carried out studies into the optimisation of processes. Such information could be combined to give the optimum design with the optimum combination of processes so that the spring size could be reduced.

The ability to predict, with accuracy, the design parameters of variable rate springs was an innovation which was expected to be particularly valuable in the North American market. The valve springs which had been used in the large and slow-revving North American engines had been lowly stressed and made from carbon steel. A separate friction damper sleeve had been used to dampen the natural spring surge. With higher engine revolutions neither carbon steel nor friction dampers would be effective and a variable rate spring, made from alloy steel, would be required.

CS7.2.2 Development of an auto-line

TSCo considered that an automated valve spring production line was needed. This line would make springs in long production runs using oil-tempered alloy wire manufactured to close tolerances. Such a manufacturing facility would effectively

Product – engine valve springs **C**

- Ability to fit springs into a smaller space and to reduce weight
- Ability to retain designed load characteristics for increasing engine life
- Ability to withstand surge at increased revolutions per minute
- Minimal operating noise
- Single spring solutions to variable rate requirements. This particularly applied to the market in North America
- Close tolerance to allow for robotic assembly
- Reduction in selling price
- Quick and efficient prototype and sample service

Figure CS7.1 Developments in valve springs

produce the designs which would satisfy the developments shown in Figure CS7.1.

CS7.2.3 Decision to install the auto-line

This line had been under consideration for some five years. The managing director had visited Japan in 1980 to see such a line. However, the decision had been delayed due in part to the cost of investment but also due to organisational and workforce considerations.

During 1982 TSCo was reorganised into five divisions. One of these was the valve spring division. This reorganisation focused attention on the process and allowed the divisional manager to prepare a specification and layout for an auto-line.

It was important to have a team approach to the introduction of an auto-line. In 1982 the union continued to adopt a militant stance and relations between management and unions deteriorated. A point at issue was the introduction of new manning levels required by the reduced demand for company products. Confrontation took place resulting in a strike. The resolution of the strike gave rise to a much improved atmosphere which, it was felt, would give the co-operation and teamwork needed for the introduction of the auto-line.

The competitive position of TSCo was being eroded. Competitors were manufacturing with reducing cost and improving quality. As a result installation and commissioning of the auto-line became a matter of great urgency. A decision was made to install and commission an automated line as a result of the strategy review made of the period 1985 to 1989. The plant was to be operational in 1985.

CS7.2.4 Quality

Statistical process control (SPC) studies to ensure the capability of this line had been carried out. By the time the divisional strategy was evolved in 1985 quality systems had become a mandatory requirement of all significant automotive manufacturers. The pressure from the automotive industry on its suppliers was led by Ford and the introduction of its Q101 system. Q101 was based on:

- Advanced quality planning.
- Quality execution.
- Performance monitoring.

Case Study 7: The Tempered Spring Company Limited

SPC underpinned all three. TSCo expected that it would be possible to reduce load testing to an 'audit' operation when the SPC studies, and the action prompted by them, were completed. It was also anticipated that this would be the main thrust in obtaining Q101 approval.

CS7.3 The total market and TSCo's share

The divisional strategy considered that increasing use of four valves or more per cylinder to improve aspiration, economy and emissions would increase the world demand for valve springs over the next five years. A four-valve arrangement allows the engine to so improve fuel mixture and burning that it performs as well as a turbo-charged version and achieves maximum combustion to minimise the emission of unburned gases. A conservative estimate was that 20 per cent of engines made in 1989 would have more than two valves per cylinder.

The need to develop engines capable of higher revolutions was expected to cause North American manufacturers to buy from European spring makers in order to acquire the proven design technology described above. This situation was expected to be short-lived. Once designs were established it was expected that North American engine builders would look for a domestic source of supply. To retain a position in the American market in the longer term TSCo would need to manufacture in North America.

The major process opportunity was seen to be the final commissioning of the fully automated valve spring production line which was expected to improve quality and production.

The 1985 divisional strategy was based on two valves per cylinder. Table CS7.1 shows the expected result of the market using two valves per cylinder and a forecast based on 20 per cent of the market using four valves per cylinder. The actual size of the market in 1989 was greater than the second forecast.

As the second spring manufacturer in Europe TSCo had a good reputation for product design and the potential, with the newly installed plant, of low-cost manufacture. Its limitation on the provision of samples was being overcome by the transfer of sample production from the bulk plant and onto other plant which was more readily available for the manufacture of samples and prototypes.

The growth in market share projected for the company was based upon an assessment of competitors' strengths and weaknesses. The new auto-line was

Table CS7.1 TSCo's share of the valve spring market

	Europe			North America		
	1985 actual	1989 forecast millions	1989 actual	1985 actual	1989 forecast millions	1989 actual
Market	134	138	180	145	133	170
TSCo forecast	16	22	20	NIL	5	4
TSCo share of Europe	11.9%		11.1%			
Forecast with 20% of the market using four valves per cylinder						
Market		156			160	
TSCo		26			6	

expected to give TSCo a price and quality advantage which would enable the forecast volume increases to be gained. The forecast target increases would result in the following sales totals:

- 1985 – actual 16 million.
- 1989 – target 27 million.

A production capacity of 27 million was anticipated after commissioning the new line.

As can be seen from Table CS7.1 TSCo did not retain its share of the European market although it was successful in its strategy for entering the American market.

CS7.4 Manufacturing

CS7.4.1 Process flow on the auto-line

A flow diagram of the auto-line is shown in Figure CS7.2. The process route is described below.

Wire stock

Oil-tempered wire purchased from a supplier in Sweden is used on the automatic line. The Swedish supplier was chosen after attempts to source in the UK had failed. The problem with UK supply was lack of consistent quality. The use of SPC on the coiling station showed up this problem. The Swedish supply is consistent in quality.

On the batch line oil-tempered wire and hard drawn wire is used. The latter is variable in quality but is cheaper. The customers who specify the product in this quality generally have smaller quantities. Thus the automatic line is dealing with high volume orders, made from a consistent raw material, while the batch line makes a wide variety of products and uses two raw materials.

Attempts have been and continue to be made to obtain a more consistent quality of hard drawn wire without success.

Coil and shape

The wire coiling and shaping machine is manually set. Numerically controlled machines are available but they are considered to have disadvantages:

- They run at a slower production rate.
- It is not possible manually to override the numerical controls in the event of software failure.

Setting takes approximately four hours and is out of sequence with setting on the grinder which takes approximately five hours. All machines are set by operators.

Inspection is manual using automatic measuring instruments. Data capture is manual.

C *Case Study 7: The Tempered Spring Company Limited*

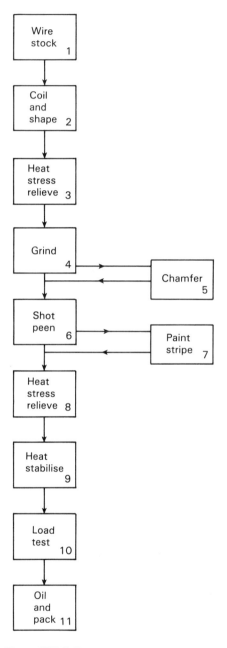

Figure CS7.2 Process route on the auto-line

Heat stress relieve

Stress relieving is carried out in a continuous feed oven. The springs are fed in 'sausages' into metre-long tubes which are taken through the furnace by conveyor. The furnace is time and temperature controlled so that the springs are warmed up for ten minutes and are then in a temperature of 400°C for twenty minutes.

Grind

The grinder is a hand-fed machine which has been converted to automatic feed. The original design for the automatic line included two grinding machines but capital availability limited the line to one grinder. Technical improvements were expected to raise the output of the grinder to that of the line but this has not been achieved and the grinder is a bottleneck.

In the flow line there is a by-pass in front of the grinder. When springs fed from the coiling machine exceed the rate at which the grinding machine can cope the springs are put into bins and ground on the batch line. The springs are subsequently fed back into the auto-line.

Chamfer

Chamfer was originally to have been incorporated into the flow line but with only some 2 per cent subjected to this operation it is more economical to chamfer on the batch line.

Shot peen

Shot peening is still carried out in batches on the old line. There is an automatic shot peener but it has not been possible to commission it together with satisfactory automatic feed and discharge. The experience gained with the automatic shot peener shows the advantages of shot peening individual springs compared with batch processing. However, the difficulties encountered with the machine have led to a decision to abandon further work until line renewal.

Paint stripe

Paint stripe continues to be a batch operation.

Heat stress relieve

The heat stress relieve process is automatic but is hand-fed. There was an automatic feeder available but following prolonged commissioning problems the same decision was made as for the shot peener.

Heat stabilise

There is automatic feed to heat stabilise.

Case Study 7: The Tempered Spring Company Limited

Load test

Load test is automatically fed. Five quality categories are produced:
- Perfect.
- Two categories which can be reworked.
- Two categories which are scrap.

Oil and pack

Oil and pack were to have been automated but the automation was abandoned due to lack of finance.

CS7.4.2 Effect of auto-line on quality

The auto-line produces higher quality because each spring receives individual treatment (with the exception of shot peening). As a result there is a reduction in the variation of quality. Batch treatment gives rise to variability, e.g. a basket of springs in a stress relieving furnace has springs on top, in the centre of the basket and at the bottom which all receive differing treatments.

CS7.4.3 Handling springs on the auto-line

Springs are difficult to handle by their nature. They have potential energy which can be released if they are dropped and they 'tangle' readily. This has limited the use of many conventional feeding devices as the line has been automated. A major factor in the success of the automated line has been the resolution of the problem of handling and feeding. Many pieces of equipment in the auto-line are identical to those in the batch process, e.g. the coiling and grinding machines but springs pass through these operations individually not in batches.

CS7.5 Commissioning and operating the auto-line

There were two important aspects to commissioning and operating the auto-line:
1. Staffing.
2. SPC.

CS7.5.1 Staffing

Volunteers were called for from throughout the factory. Preference was given to operatives with spring experience. More volunteers were obtained than were required. An initial screen was programmable logic control (PLC) training. This was arranged with the equipment suppliers and resulted in a number of the volunteers wishing to withdraw their application.

Training was to be a mixture of 'on and off the job'. Following the PLC training the operatives joined the commissioning engineers to gain experience in operating the plant.

Commissioning and operating the auto-line

Maintenance was not provided on a 24-hour basis although the plant was to operate on three shifts. Operatives had to undertake maintenance as part of their role. This consisted of routine maintenance such as cleaning and oiling. There were also blockages caused by springs tangling. Operatives had to be trained to remove guards, clear blockages and put the plant back together.

As the team evolved it was found that four not five operatives were required. Natural leaders emerged who were recognised by the management of the valve spring division although there was no financial reward.

CS7.5.2 Statistical process control

Management decided to adopt a manual rather than a computerised system of SPC. The result has been that operatives have 'ownership' of the system. They became interested in the control charts and understood that they could determine if a process was going out of control. They felt able to contribute from their expertise to solving the problems which arose and to keep the process in control.

The reduction in rework shown in Figure CS7.3 was the result of bringing the process under control by the application of SPC. A vital part of this was bringing raw material supplies under control. Operatives have formed improvement teams which are supported by management.

An example of the need of management support in the application of SPC can be seen in the case of the operator on the shot peen process who had taken great care to set up the operation. The control charts kept showing the process going out of control. Management worked with the operative on the problem which was finally identified as the quality of the shot.

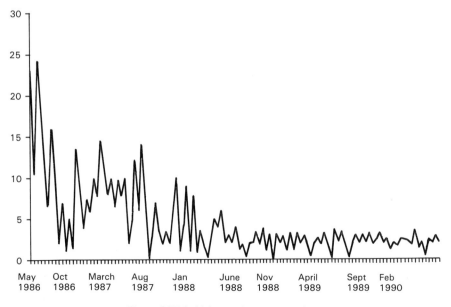

Figure CS7.3 Valve spring rework chart

Case Study 7: The Tempered Spring Company Limited

CS7.6 Success of the divisional strategy

CS7.6.1 Market

As already stated, the share of the European market had dropped. Nevertheless, TSCo had successfully entered the American market. Product quality had been improved and cost lowered as a result of building the auto-line. The problems involved with commissioning the auto-line could not be solved on the British plant until it was renewed. However, the experience gained was the basis for the new plant to be opened in America with a factory that has space to install six auto-lines.

CS7.6.2 Production efficiency

Table CS7.2 shows the production in October 1983 from the batch plant in comparison with October 1988 when both the batch and auto-line were in operation. The productivity per operative has increased by 54 per cent and the rework has dropped from one-third of output to a level consistently below 3 per cent. This can no longer be improved by SPC but requires plant with an enhanced capability.

CS7.6.3 Production cost

The savings that had been made by the introduction of the auto-line can be obtained by analysing what would have been the cost of producing the 1988 production with the 1983 batch facility.

- Operative savings. The improved productivity of 54 per cent means that 49 operatives would have been required in 1983 to do the work of 32 operatives in 1988. There is a saving of 18 operatives which, at an oncost of £15,000 per operative per year, gives an annual saving of £225,000.
- Rework saving. There is a saving due to the reduction in the quantity of rework. A normal year would equate to twelve 20-day periods. Thus October 1988 production in 20 days would be 2,135,200 springs. This would give a monthly saving in rework of:

$$\frac{213520 \times 80\% \text{ hot pressed} \times 32\% \text{ rework saving} \times £26.1 \text{ CCR}}{2400}$$

$$= £5940/\text{mnth}$$

Table CS7.2 Comparison of batch and auto-line manufacture

	October 1983 Batch	October 1988 Batch + auto
Output of springs	1,776,000	2,660,000
Days worked	20	25
Direct operatives	41	32
Springs/operative/day	2,165	3,337
Percentage rework	35	3

Success of the divisional strategy

Only hot pressed springs are reworked and this amounts to 80 per cent of spring manufacture; CCR = cost centre rate.

The monthly saving in rework would amount to £71,330 per year. Total saving per year for auto-line = £326,330.

CS7.6.4 Capital expenditure on the auto-line

- The capital cost of the project at £646,500 showed an overspend of 20 per cent.
- The actual level of production in August/September 1988 was 2,400 springs per hour. The planned output was for 3,870 springs per hour. The main reason for the difference was that the line was planned with two grinding machines but only one was installed to save capital expenditure.
- There were four operatives per shift compared with the planned five per shift.

However, the savings of £326,330 shown above would give a 2.2 year payback at an annual interest rate of 12 per cent.

Abbreviations

ABC	activity-based costing
AGV	automated guided vehicle
AIMS	advanced integrated manufacturing system
AMT	advanced manufacturing technology
AOP	annual operating plan
BA	business area
BL	British Leyland
BMC	British Motor Corporation
BOM	bill of materials
BS	British Standard
CAD	computer-aided design
CAM	computer-aided manufacture
CBI	Confederation of British Industry
CCR	cost centre rate
CCS	central control system
CEO	chief executive officer
CIM	computer-integrated manufacture
CLT	cumulative leadtime
CNC	computer numerically controlled
CWQC	company-wide quality control
DCF	discounted cash flow
DFM	design for manufacture
DNC	direct numerical control
DTI	Department of Trade and Industry
EBQ	economic batch quantities
E-CLT	effective cumulative leadtime
EDI	electronic data interchange
EOQ	economic order quantities
EU	European Union
FAS	fixed assembly schedule
FMS	flexible manufacturing system
GATT	General Agreement on Tariffs and Trade
GDP	gross domestic product
HRM	human resource management
IC	integrated circuits

Iip	Investors in People scheme
IR	industrial relations
JIT	just-in-time
KRA	key result area
LPA	loss prevention audit
MBO	management by objectives
MCO	manufacture to customer order
MIO	management improvement opportunities
MLT	manufacturing leadtime
MPC	manufacturing planning and control
MPS	master production schedule
MRP	materials requirement planning
MRPII	manufacturing resource planning
NAFTA	North American Free Trade Association
NC	numerically controlled
N-CLT	native-cumulative leadtime
NIC	newly industrialised country
NIESR	National Institute of Economic and Social Research
NVQ	national vocational qualification
OECD	Organisation for Economic Co-operation and Development
OPT	optimised production technology
PBR	payment by results
PCB	printed circuit board
PDS	product design specification
PLC	programmable logic controller
QC	quantity circle
QFD	quality function development
ROC	return on capital
ROI	return on investment
SAM	system for Annesley manufacturing
S&OP	sales and operations planning
SBU	strategic business unit
SE	simultaneous engineering
SMT	surface mount technology
SOP	sales order processing
SPC	statistical process control
SQ	strategic quantification
SQC	statistical quality control
SWOT	strengths, weaknesses, opportunities and threats
TCS	transport control system
TDC	total departmental cost
TEC	Training and Enterprise Council
TFC	total factory cost
TQM	total quality management
TUC	Trades Union Congress
VOC	voice of the customer
WIP	work in progress

Bibliography

Abernathy, W.J. (1978) *The Productivity Dilemma*, Johns Hopkins University Press.
The Acard Report (1983) HMSO.
Barney, J.B. (1986) Organizational culture: can it be a source of sustained competitive advantage?, *Academy of Management Review*, Vol. 11, No. 3.
Bellis-Jones, R. (1992) Activity-based costing, in *Management Accounting Handbook*, ed. C. Drury, Butterworth Heinemann, in association with the Chartered Institute of Management Accountants.
Bentley, J. (1991) Integrating design and manufacturing strategies for business information *International Journal of Technology Management*, Vol. 6, Nos 3/4.
Bhimani, A. and Bromwich, M. (1992) Management accounting: evolution in progress, in *Management Accounting Handbook*, ed. C. Drury, Butterworth Heinemann, in association with the Chartered Institute of Management Accountants.
Bhimani, A. and Pigott, D. (1992) Implementing ABC: a case study of organizational and behavioural consequences, *Management Accounting Research*, Vol. 3.
Bolwijn, P.T. and Kumpe, T. (1990) Manufacturing in the 1990s – productivity, flexibility and innovation, *Long Range Planning*, Vol. 23, No. 4.
Bromwich, M. (1990) The case for strategic management accounting: the role of accounting information for strategy in competitive markets, *Accounting, Organization and Society*, Vol. 15, No 1/2.
Bromwich, M. and Bhimani, A. (1989) *Management Accounting: Evolution not revolution*, CIMA.
Broughton, T. (1990) Simultaneous engineering in aero gas turbine design and manufacture, *Proceedings of the 1st International Conference, Simultaneous Engineering*, Status Meetings Ltd.
Butcher, M.C. (1986) Integrated manufacture – from concept to reality, *IMechE Conference Publication*, 1986 – 10.
Butcher, M.C., Perkins, N.R., Wells, K.E. and Peto, M.G. (1987) An advanced integrated manufacturing system for turbine and compressor discs, *Proceedings of The Institution of Mechanical Engineers*, Vol. 201, No. B3.
Cam-I (1988a) *Management Accounting in Advanced Manufacturing Environments. A Survey*, prepared by Coopers & Lybrand, Ernst & Whinney and Peat Marwick McLintock, January.
Cam-I (1988b) *Cost Management for Today's Advanced Manufacturing: The CAM-I conceptual design*, Harvard Business School Press, Computer Aided Manufacturing–International Inc.

Carter, D. and Stilwell Barker, B. (1991) *Concurrent Engineering: The product development environment for the 1990s*, Addison-Wesley.
Claret, J. (1987) Accounting for AMT, *Industrial Computing*, May.
Cooper, R. and Kaplan, R.S. (1988) Measure costs right: make the right decisions, *Harvard Business Review*, September–October.
Coyle, D. (1992) Slow burn for learning, *Management Today*, December.
Crosby, P.B. (1989) *Quality Is Free*, McGraw Hill.
Dale, B.G. and Plunkett, J.J. (1991) *Quality Costing*, Chapman & Hall, London.
De Meyer, A., Nakane, J., Miller, J.G., and Ferdows, K. (1987) Flexibility: the next competitive battle, Publication 86/31, INSEAD working paper, March.
Department of Employment and Productivity (1969) *In place of strife: A policy for industrial relations*, Cmnd 3888, HMSO, London, June.
Department of Trade and Industry (1991), *The Investment in Britain Bureau*, HMSO, October.
De Vries, J. and Rodgers, L. (1991) Bridging business boundaries, *The TQM Magazine*, IFS (Publications) Ltd.
Donovan Commission (Royal Commission on Trade Unions and Employees Associations) (1968) Report presented to parliament, HMSO, London, June, reprinted 1975.
Drucker, P.E. (1990) The emerging theory of manufacture, *Harvard Business Review*, May–June.
Edmonds, J. (1990) Quoted in 'The politics and complexity of trade union responses to new management practices', M.M. Lucio and S. Weston, *Human Resource Management Journal*, Vol. 2, No. 4, 1992.
Eltis, W. and Fraser, D. (1992) The contribution of Japanese industrial success to Britain and to Europe, *National Westminster Bank Quarterly Review*, November.
Financial Times, 26 March 1993, Lisa Wood.
Financial Times, 21 July 1993.
Gallie, D. and White, M. (1993) Employee commitment and skills revolution, *The Employment Survey in Britain*, Policy Studies Institute.
Garvin, D.A. (1988) *Managing Quality: The strategic and competitive edge*, Free Press, New York.
Glynn, S. and Gospel, H. (1993) Britain's low skill equilibrium: a problem of demand?, *Industrial Relations Journal*, Vol. 24, No. 2.
Goldhar, J.H. and Jelinek, M. (1983) Plan for economies of scope, *Harvard Business Review*, November–December.
Hall, R.W., Johnson, H.T. and Turney, P.B.B. (1991) *Measuring Up: Charting pathways to manufacturing excellence*, Business One Irwin, Homewood, Illinois.
Hamel, G. and Prahalad, C.K. (1989) Strategic intent, *Harvard Business Review*, May–June.
Hayes, R.H. and Wheelwright, S.C.W. (1984) *Restoring our Competitive Edge*, Wiley, USA.
Herbert, K. (1992) *Fencing with Time*, Kodak.
Hill, T.J. (1993) *Manufacturing Strategy: The strategic management of the manufacturing function*, Macmillan.
Hiromoto, T. (1988) Another hidden edge – Japanese management accounting, *Harvard Business Review*, July–August.
Hutton, W. (1991) Why Britain can't afford the City, *Management Today*, September.
Industrial Computing (1986) April, pp. 19–21.
Kano, I. (1993) pp. 10–13 in *Strategic Benchmarking*, ed. G.H. Watson, Wiley, New York.
King, D. (1987) Approaches to manufacturing, in *The Management of Manufacturing*, eds R.P. Toone and D. Jackson, IFS (Publications) Ltd.

Lucio, M.M. and Weston, S. (1992) The politics and complexity of trade union responses to new management practices, *Human Resource Management Journal*, Vol. 2, No. 4.

Mackey, J. (1991) MRP, JIT and automated manufacturing and the role of accounting in production management, in *Issues in Management Accounting*, eds D. Ashton, T. Hopper, and R.W. Scapens, Prentice Hall.

Makido, T. (1989) Recent trends in Japan's cost management practice, in *Japanese Management Accounting*, eds. Y. Monden and M. Sakurai, Productivity Press, Cambridge, Massachusetts.

Miller, J.G., De Meyer, A. and Nakane, J. (1992) *Benchmarking Global Manufacturing*, Business One Irwin, Homewood, Illinois.

Milsome, S. (1993) The impact of Japanese firms on working and employment practices in British manufacturing industry, *IRS Review and Report*, July.

Monden, Y. (1989a) Framework of the just-in-time production system, in *Japanese Management Accounting*, eds Y. Monden and M. Sakurai, Productivity Press, Cambridge, Massachusetts.

Monden, Y. (1989b) Cost accounting and control in the just-in-time production system: the Daihatsu Kogyo experience, in *Japanese Management Accounting*, eds Y. Monden and M. Sakurai, Productivity Press, Cambridge, Massachusetts.

Morrow, M. and Connolly, T. (1994) Practical problems of implementing ABC, *Accountancy*, January.

Mortimer, J. (ed.) (1985) *Integrated Manufacture, Ingersoll Engineers*, IFS (Publications) Ltd.

Mueller, F. (1992) Flexible working practices in engine plants: evidence from the European automobile industry, *Industrial Relations Journal*, Vol. 23, No. 3.

New York Times, October 1991.

Nichols, K. (1990) Competing through design – today's challenge, *Proceedings of the 1st International Conference, Simultaneous Engineering*, Status Meetings Ltd.

Panzar, J.C. and Willig, R.D. (1981) Economies of scope, *American Economic Review*, Vol. 71, No. 2.

Parnaby, J. (1988) Creating a competitive manufacturing strategy, *Production Engineer*, July–August.

Peters, T.J. and Waterman, R.H. (1982) *In Search of Excellence*, Harper & Row, New York.

Porter, M.E. (1980) *Competitive Strategy*, Free Press, New York.

Porter, M.E. (1985a) Technology and competitive advantage, *Journal of Business Strategy*, Winter, Vol. 5, No. 3.

Porter, M.E. (1985b) *Competitive Advantage*, Free Press, New York.

Prahalad, C.K. and Hamel, G. (1990) The core competence of the corporation, *Harvard Business Review*, May–June.

Primrose, P.L. (1988) AMT investment and costing systems, *Management Accounting*, October.

Pugh, S. (1991) *Total Design*, Addison-Wesley.

Puttick, J. (1987) 'Marketing pull – manufacturing push' springboard for competitive advantage, in *The Management of Manufacturing: The competitive edge*, eds R. Toone and D. Jackson, IFS (Publications) Ltd.

Quinn, J.J. (1985) How companies keep abreast of technological change, *Long Range Planning*, Vol. 18, No. 2.

Radford, G.D. (1989) How Sumitomo transformed Dunlop tyres, *Long Range Planning*, Vol. 22, No. 3.

Rhefeld, J.E. (1990) What working for a Japanese company taught me, *Harvard Business Review*, November–December.

Salamon, M. (1987) *Industrial Relations: Theory and practice*, Prentice Hall.

Sanderson, M. (1992) *BSI News*, November.
Shingo, S. (1983) *A Revolution in Manufacturing: The SMED system*, Japan Management Association (English edition: Productivity Press, 1985).
Shingo, S. (1986a) *The Sayings of Shigeo Shingo: Key strategies for plant improvement*, Nikkan Kogyo Shimbum, Ltd (English edition: Productivity Press, 1987).
Shingo, S. (1986b) *Zero Quality Control: Source inspection and the poka-yoke system*, Productivity Press.
Shingo, S. (1988) *Non-stock Production: The Shingo system for continuous improvement*, Productivity Press.
Simmonds, K. (1992) Strategic management accounting: what makes it different?, *Manufacturing Technology International*.
Skinner, W. (1974) The focused factory, *Harvard Business Review*, May–June.
Slack, N. (1991) *The Manufacturing Advantage*, Mercury Books.
Smith, A. (1776) *The Wealth of Nations*.
Smith, P.G. and Reinertsen, D.G. (1991) *Developing Products in Half the Time*, Van Nostrand & Reinhold, New York.
Stalk, G., Evans, P. and Shulman, L.E. (1992) Competing capabilities: the new rules of corporate strategy, *Harvard Business Review*, March–April.
Sunday Times, 9 January 1994.
Tanaka, M. (1989) Cost planning and control systems in the design phase of a new product, in *Japanese Management Accounting*, eds Y. Monden and M. Sakurai, Productivity Press, Cambridge, Massachusetts.
Taylor, F.W. (1911) Scientific management, comprising shop management. The principles of scientific management, Testimony before the Special House Committee, Harper & Bros, New York.
Taylor, W. (1991) The logic of global business: an interview with ABB's Percy Barnevik, *Harvard Business Review*, March–April.
Thomas, P. (1987) Reducing costs by raising quality, in *The Management of Manufacturing: The competitive edge*, eds R. Toone and D. Jackson, IFS (Publications) Ltd.
Tichy, N. and Charan, R. (1989) Speed, simplicity, self-confidence: an interview with Jack Welch, *Harvard Business Review*, September–October.
The Times, 12 February 1992, K. Eason.
The Times, 1 March 1992.
The Times, 14 April 1992.
The Times, 7 November 1992.
The Times, 22 November 1992.
The Times, 24 November 1992.
The Times, 24 March 1993.
The Times, 23 April 1993, On the road to discovery, K. Eason.
The Times, 19 July 1993.
The Times, 22 July 1993, A. Sykes.
The Times, 23 July 1993.
The Times, 21 December 1993.
Turney, P.B.B. (1991) Common cents, the ABC performance breakthrough, how to succeed with activity-based costing, *Cost Technology*, Hillsboro, Oregon.
Venkatesan, R. (1992) Strategic sourcing: To make or not to make, *Harvard Business Review*, November–December.
Voss, C. (1992) Applying service concepts in manufacturing, *International Journal of Production and Operations Management*, Vol. 12, No. 4.
Waldron, D. and Galloway, D. (1988) Measure your way to profit, *Certified Accountant*, May.

Walker, R. (1992) Rank Xerox–management revolution, *Long Range Planning*, Vol. 25, No. 1.
Wallace, T.F. (1985) *MRPII Making it Happen*, Oliver Wight Ltd Publications, Essex Junction, USA.
Warner, M., Wobbe, W. and Brödner, P. (eds) (1990) *New Technology and Manufacturing Management*, John Wiley.
Wickens, P. (1987) *The Road to Nissan*, Macmillan, London.
Womack, J.P., Jones, D.T. and Roos, D. (1990) *The Machine that Changed the World*, Rawson Associates, New York.

Index

advanced manufacturing technology, 25
autonomation, 43, 117

benchmarking, 100–4, 106–113, 124
  see also strategic bench marking
brownfield operations, 19, 20–1, 33, 34, 100, 140, 146, 152
BS5750/ISO9000, 98, 121

cells, 76, 86, 91–2
  at Stanley Tools, 249–251
change management, 16–18, 19
competition
  competing on flexibility, 38, 44–6
  competing on innovation, 38, 46–7
  competing on price/cost, 38, 42–3, 53–4
  competing on quality, 38, 43–4
  competing on service, 47–8
  competitive advantage, 6, 13–14, 49, 50, 92, 110
  competitive criteria, 37–48, 71–2
  global competition, 9–11
complexity, 70–1
computer integrated manufacture, 25–6, 89
consumer durables, 15, 37, 40, 42, 51–2, 113, 134
core competence, 37, 67, 69–70
cost, 53–4
  control, 35–6, 156–8
  non-financial measures, 165
  reduction, 42, 54, 156, 157
  value chain, 164–5
costing systems
  activity-based costing, 158–60

adequacy of systems, 155
back flush relief of inventory, 163–4
budgets and budgetary control, 157
cost modelling/simulation, 160
cost of quality, 160
life cycle costing, 164
strategic management accounting, 161–2
target costing, 115–16, 156, 160–1
throughput accounting, 162–3
culture, 14, 21, 43, 124
  customer, 14, 21, 37, 47–8, 74, 93, 100–4, 108, 111–12, 138

design for manufacture, 21, 40–1, 55–6, 90–1, 113–16
discounted cash flow, 152

economic factors in the environment, 1–9
economies of scale, 42
economies of scope, 45
efficiency, 42, 54–5
electronic data interchange, 9–10, 28, 30, 60, 86

finance, 23
  capital investment
    short termism, 35, 59, 150–2
  decision making, 165–6
  financial viability, 33–5
  investment appraisal, 35–7, 152–5
  investment, tangible and intangible benefits, 34–6, 152–4
  investment decisions, 61
  provision of finance, 150–2

271

Index

flexibility, 45–6
  see also competition – competing on flexibility; human resources flexibility; process flexibility
focus, 50, 55, 58, 59, 68, 70–2, 92, 104–5
functional boundaries, 21, 50, 54–5, 92

greenfield operations, 19–20, 33–5, 140, 146, 152–4

human resources, 53, 125–7
  appraisal and reward, 145–8
  as a competitive factor, 125–9
  decisions, 61
  development of, 142–4
  effects of new technology on *see* technology
  empowerment, 63, 86, 88–9, 105, 119–20, 140
  flexibility, 147–8
human resource management, 30, 61, 125–32, 136–41, 148
  criticism of new working practices, 141

industrial relations, 20–1, 31–2, 132
innovation, 38, 40, 46–7
integration, 21–6, 44, 50, 56–9, 92–3, 104–5, 138
  horizontal, 89–91
  process, 73–6
  supply chain, 77–86
  vertical, 86–9
just-in-time, 15–16, 25, 29, 36, 46, 75, 82–3, 106, 116, 125, 140–1

Kanban *see* manufacturing planning and control

lean production, 14–16, 19, 56–7
learning/experience curve, 41, 42, 53–4
leadtime, 26, 72, 85
life cycle
  costs, 40–1
  product life cycle, 38–42, 55–6, 113

management philosophy, 17–18
manufacturing, 21–2, 58
  capability, 41, 72, 75–6
  development of Western manufacturing, 52–6
  impact of Japanese manufacturing in UK, 51–2

importance of UK manufacturing, 1–6, 12
productivity, 6–9, 30, 33, 52, 127
strategy, 23, 34, 37, 58, 59, 61, 92
manufacturing planning and control, 77, 78–9, 83, 84–5
  class A MRPII user definition, 218–9
  kanban, 25, 46, 83, 118
  at Kodak Ltd, 211–18
  MRP/MRPII, 17–18, 25, 79–81, 130
  network analysis, 79
  OPT, 81–2
  at Stanley Tools, 247–9
  systems, 78–9
  time fences in MRPII, 220–7
manufacturing systems engineering, 91–2
market entry on time, 56
marketing, 21–2, 55
marketing strategy, 23, 92
  at Stanley tools, 244–7

objective setting, 87–8
operational pressures, 19, 20, 59 70, 153, 154
order qualifiers/winners, 37, 47–8
organisation
  functional, 54–5, 104–5
  structure, 59–61, 63–6, 92

payment by results, 53, 127–8
  at Stanley Tools, 251–3
price/product cost, 38, 42–3, 122, 156
process capability, 113–14, 116, 120
process flexibility, 26

quality, 38, 43–4, 91, 94
  continuous improvement, 15, 31, 49, 122
  cost of quality, 95, 98, 122
  product quality and process capability, 113–17
  quality circles, 117–19
  quality function deployment, 100–4, 106
  quality planning, 104–7
  right first time/zero defects, 26, 84, 101, 116–17
  SPC at TSCo, 255, 261
  standards, 98–9
  statistical process control/statistical quality control, 98–9, 109, 116, 118

quality (*continued*)
   supplier quality, 120–1
   total quality management, 15–16,
      43–4, 96–7, 104–5, 124, 140–1

research and development, 23, 46

service, 47–8
set-up time reduction, 26, 57, 76
   at DSF refractories, 182–3
simplification, 25, 74–5, 83
simulation, 34
simultaneous engineering, 56, 89, 105
   development at Rolls-Royce, 234–42
stock, 77–8
strategic bench marking, 67–8
strategic business units, 21, 28, 48,
   63–6, 69, 92, 181–2
strategic intent, 66–7
strategy, 21–3, 50, 61–70, 92
   Fisher Controls
      company strategy, 186–7
      manufacturing strategy, 187
   Hepworth
      CIM strategy, 201–10
      entry strategy, 196–8
   Rolls-Royce
      AIMS strategy, 231–3
      manufacturing strategy, 228–231
strategy of the integrated manufacturing
   company, 58–62
success, 13–14, 19, 153
supplier partnering, 29, 83–4,
   121–2
supply chain, 46, 50

management, 27–30, 77, 83–6, 101,
   120–2, 138

team working, 21, 33, 63, 86, 120, 136,
   147, 148
technology
   cost control of new technology, 35–6
   development and R&D, 23, 39, 46–7
   investment in new technology, 32–5
   labour and new technology, 30–1,
      135–6
   of manufacture, 23–6, 137
   new working practices, 138
   strategy, 47, 68–70
tools and techniques, 14
   experience with their introduction,
      18–20
total quality management *see* quality
Toyota production system, 15–16,
   56–7, 131–2
trade offs, 55, 68
trade unions, 54–5 132–6
   response to HRM, 138–9
   changing role, 132–4
   new technology effects, 31–2
training and development, 92, 138,
   142–4

uncertainty, 45, 70–1

value added, 74, 76, 86

waste elimination, 15, 16, 36, 57, 79,
   82, 131
work flow organisation, 74–5, 91